A Field Guide to

Frogs

of Australia

From Port Augusta to Fraser Island including Tasmania

Martyn Robinson

AUSTRALIAN
MUSEUM

An Australian Museum/Reed New Holland Publication

ACKNOWLEDGEMENTS

No work is ever produced in a vacuum, and the efforts, advice and assistance of the following Australian Museum staff are gratefully acknowledged:

Ms E. Cameron	Research Officer
Dr H. Cogger	Deputy Director
Ms F. Fletcher	Typist
Dr A. Greer	Senior Research Scientist
Mr P. Macinnis	Education Officer
Ms M. Martin	Manager, Education Programs

Most of the photographs used in the book were taken by Dr Cogger, but the following photographers need also to be thanked: D. & V. Blagden, H. Ehmann, P. German, G Grigg, G.A. and M.M. Hoye, R.W.G. Jenkins, G. Little, M. Littlejohn, M. Mahony, P. Roach, G.E. Schmida, M.G. Swan, R. Whitford, J. Wombey. Their photos are acknowledged in the text. Thanks also to the staff of the National Photographic Index.

The illustrations have been prepared by the author.

I would also like to thank those outside the Museum who checked the manuscript for errors, particularly Dr Margaret Davies and Dr Mike Mahony, and Dr G. Ingram for his rendition of the call of *Rheobatrachus silus* over the telephone.

Thanks, too, to Peter Macinnis who typed and did a preliminary edit of the manuscript.

My special thanks to Lynne McNairn, who put up with me during the writing of this work.

Published in 1998 by Reed New Holland
an imprint of New Holland Publishers (Australia) Pty Ltd
Sydney • Auckland • London • Cape Town

1/66 Gibbes Street Chatswood NSW 2067 Australia
218 Lake Road Northcote Auckland 0627 New Zealand
86 Edgware Road London W2 2EA United Kingdom
80 McKenzie Street Cape Town 8001 South Africa

First published in Australia by Reed Books in 1993

Reprinted by Reed New Holland in 1998, 1999, 2000, 2002, 2004, 2005, 2007.

National Library of Australia
Cataloguing-in-Publication Data:

Robinson, Martyn.
A field guide to frogs of Australia : from Port Augusta to Fraser Island including Tasmania.
Includes index.
ISBN 9781876334833 (pbk.).
1. Frogs - Australia - Identification. I. Title.
597.890994

Publisher: Lousie Egerton
Editor: Rosemary Neilsen
Designer: Robert G. Taylor
Cover Design: Nanette Backhouse
Printer: Hong Kong Graphics & Printing Ltd

FRONT COVER PHOTOGRAPHS: Top left: Green and Golden Bell Frog (*Litoria aurea*), Ken Griffiths/Nature Focus; Centre left: Holy Cross Toad (*Notaden bennetti*), Marion Anstis; Bottom left: Heath Frog (*Litoria littlejohni*), John Wombey; Top right: Red-eyed Tree Frog (*Litoria chloris*), Marion Anstis; Cut out: Northern Corroboree Frog (*Pseudophryne pengilleyi*), Marion Ans

BACK COVER PHOTOGRAPHS: Top left: Painted Fro (*Neobatrachus pictus*), Marion Anstis; Bottom left: Tasmanian Tree Frog (*Litoria burrowsae*), Paul Swiatkowski; Top right: Dainty Tree Frog (*Litoria gracil ta*), Marion Anstis; Bottom right: Blue Mountains Tree Frog (*Litoria citropa*), Marion Anstis

CONTENTS

FOREWORD

It is extremely important for us to know more about the health of our frog populations. Some species and populations of Australian frogs appear to be declining and we cannot afford to lose any of them.

Why, you may ask. Frogs are a very interesting group of animals in their own right — they have been around for at least 180 million years, they breed in a variety of fascinating and bizarre ways, and they make a major contribution to our own well-being.

Not only are frogs important elements in many food chains valued by humans (for example, those of wetlands), but they are also good indicators of the health of the environment — especially of water quality and the levels of pesticides and other toxins present in various ecosystems.

Perhaps most importantly, they are delightful creatures to know. Once you have experienced the sound of a full-throated breeding chorus of frogs on a warm spring evening, I believe you'll never again take these little amphibians for granted.

For your frog records to be really effective, correct identification of any of the 90-odd frogs of south-eastern Australia is essential. This field guide by Martyn Robinson should make the task not only much easier, but also much more accurate. Keys, descriptions and illustrations combine to characterise the species you may find in your region and provide information that could initiate a life-long interest in frogs.

Hal Cogger
Deputy Director, Australian Museum

ABOUT THIS BOOK

The area covered by this book is shown in the map at the front of this book. Essentially, it is the area encompassing south-eastern Queensland (including Fraser Island), all New South Wales, all Victoria, and all Tasmania, as well as the south-eastern corner of South Australia, across as far as Port Augusta.

This area covers a wide range of environments, from alpine areas to arid zones, from rainforests to coastal heath, and from pristine forests to urban backyards. In short, it covers any area where frogs may be found.

This field guide has been expanded from an original guide to New South Wales frogs that was developed for the Frog Watch program in that State. Frog Watch is a frog monitoring program involving schools, amateur herpetologists and anyone who wants to be involved in assessing the status of frogs in New South Wales. It is run from the Australian Museum. This expansion of the original guide was motivated by requests from people living just outside of New South Wales and by the realisation that some of the frogs recorded just beyond New South Wales may also be found inside the State's borders. The wider coverage will hopefully prove useful to many more people than originally. Species' distributions are shown across Australia, rather than just in the study area.

Other authors involved in the study of frogs will occasionally be found to use different terms or names. Until a general consensus is reached, the names of the species, and the terms used in this book follow the lead given in Cogger (*Reptiles and Amphibians of Australia,* 5th ed. published by Reed Books, 1992).

FROG CONSERVATION

In the late seventies and early eighties, scientists around the world noticed a sudden decline in the populations of frogs and other amphibians. At first, they thought these reductions were local variations, of little significance. It was only after the information was pooled that concern spread. In some cases, the cause of the decline was obvious: things like drained breeding sites, changes in land use, acid rain, drought, salination and pesticides. Each of these elements has either killed frogs or tadpoles, or made it harder for frogs to breed.

What was harder to explain was the decline and even apparent extinction of frog species in seemingly pristine natural settings in nature reserves and national parks. Information is now being gathered by Frog Watch programs in several states to enable us to build up a picture of the status of various frog species and populations. We will be able to see where they are declining and where some species seem to be spreading in range or increasing in number.

Beyond this study, there are several other things we can do to maintain and even boost frog numbers. Don't drain breeding sites of water. Don't introduce new fish species to ponds and creeks. Don't use pesticides if you can help it (frogs do a good job of eating many insect pests). Don't collect frogs or tadpoles from national parks. The absence of suitable breeding sites may be a limiting factor for some species — a backyard fish pond without fish will often be colonised by local frog species.If mosquitoes are a problem, a few small, "gentle", fish will eat them without worrying the tadpoles beyond hatchling stage. The White Cloud or Mountain Minnow (*Tanicthys albonubes*) is ideal, and cheaply available from most aquarium shops. Do not introduce this or any other fish into natural water bodies — we have enough introduced fish pests as it is!

You can even design a pond that will suit one group of frogs over another. A raised or free-standing pond will be more accessible to tree frogs, a sunken pond will suit ground-dwelling frogs, and a pond with reeds and other vegetation growing out of it will suit swamp-dwelling species. In most cases, frogs will rapidly find and colonise the appropriate pond. If a delay occurs or local frogs have been severely reduced, then some locally caught tadpoles can be added. Don't introduce frog species from other areas; it may cause severe problems for local species. Before you introduce any tadpoles to your pond, check with your local wildlife authorities, as most states have some level of protection on frogs and tadpoles — you may need to obtain a permit.

When tadpoles grow legs and begin to resorb their tails, the grounds around the pond should be well watered, to boost their survival rate. Above all, keep records of tadpoles' progress, as the information will be of value in scientific research.

FROG CLASSIFICATION

Using the familiar Green Tree Frog as an example, we shall explore how and why amphibians are classified. This section also introduces some of the technical terms you will need to know. The names used for a few classifications remain controversial and may not be agreed to by all scientists working in the field. In all cases, I have followed the preferences of Cogger as shown in his fifth edition of *Reptiles and Amphibians of Australia.*.

Frogs are amphibians. All species of amphibians are vertebrate animals (i.e. they have a backbone) and are dependent on outside temperature for their body heat (i.e. they are cold-blooded or ectothermic). They have soft skins, with a variety of glands which keep the skin moist, produce toxic secretions to deter predators, or provide a variety of other functions. Where limbs are present, they have typically four toes on the front limbs, and five on the hind ones. Most species require damp conditions or water to breed, and have gills in the larval stage. As adults, they breathe air through their lungs. The Green Tree Frog (*Litoria caerulea*) is an ectothermic vertebrate with a soft, moist, glandular skin and limbs with four fingers and five toes. It lays eggs in water, which hatch into gilled tadpoles. These later develop legs and lungs and emerge onto land.

Amphibians are divided into three orders:

1. The Caudata (or tailed amphibians), including newts and salamanders (like the "Mexican Walking Fish", sold in pet shops), which keep their tails in their adult form and whose larvae develop their front limbs first. (The other name for this group is Urodeles.) No members of this order are found naturally in Australia.

2. The Gymnophiona (or Caecilians), a small, secretive group of limbless, worm-like amphibians. Some of them may have small scales in the skin and internal fertilisation. (The other name of this group is Apoda.) No members of this order are found naturally in Australia.

3. The Salientia (or frogs and toads) as adults are tail-less amphibians with four limbs, and have mainly aquatic larvae whose hind limbs appear first. (The other name of the group is Anura.) This is the only order found naturally in Australia.

The Green Tree Frog is tail-less, has four limbs, and hind limbs are visible on the tadpole before the front limbs. It follows that the Green Tree Frog is a member of the order Salientia.

The order Salientia is divided up into families. These families contain groups of frogs that are more similar to each other than they are to frogs from other families. Three families occur in south-eastern Australia:

1. The family Bufonidae contains one introduced species, with a rough, warty skin, no teeth, no adhesive toe discs and enlarged poison glands (the Cane Toad).

2. The family Myobatrachidae contains a number of species, which have skins of various textures, lack adhesive toe discs in most species, usually have teeth and rarely climb trees.

3. The family Hylidae contains a number of species, with skins of various textures, which have adhesive toe discs in most species, have teeth, and frequently climb trees and rocks.

The Green Tree Frog has large adhesive toe discs, and frequently climbs trees. The Green Tree Frog is a member of the family Hylidae.

The family Hylidae is divided into three different genera (plural of "genus"), two of which are found in south-eastern Australia. A genus is a group of species more similar to each other than they are to members of another genus.

1. The genus *Cyclorana* includes burrowing species which do not have adhesive discs on their toes, despite their being hylid frogs (the Hylidae all have a similar skeletal arrangement and their tadpoles are a similar shape).

2. The genus *Litoria* contains species which possess adhesive toe discs (or the rudiments of them) and which range from wholly terrestrial (ground-living) to arboreal (tree-living) forms.

The Green Tree Frog possesses toe discs. The Green Tree Frog is a member of the genus *Litoria*.

Thirty-four species of *Litoria* are known to occur in south-eastern Australia. Only one is bright green with a few white flecks on the sides, with a white belly, a thick skin fold over the eardrum (tympanum), toes three-quarters webbed, fingers one-third webbed, with large adhesive toe discs. This species is *caerulea*. The Green Tree Frog has all these characteristics, so the Green Tree Frog is *Litoria caerulea*. The genus is written before the species name and always has a capital letter.

In this classification system frogs are catalogued in different groups of increasing detail the further down the classification system you go. In this way, relationships and similarities can be noted and the uniqueness of various species appreciated. Species are the building blocks of the world's fauna.

To sum up, the classification for the Green Tree Frog runs like this:

Phylum: Chordata (possessing a hollow nerve chord running down the back)
Sub-phylum: Vertebrata (...and a backbone)
Class: Amphibia
Order: Salientia/Anura
Family: Hylidae
Genus: *Litoria*
Species: *Litoria caerulea* (Green Tree Frog)

FIELD OBSERVATIONS

Frogs can be found almost anywhere in Australia, in urban, rural and bushland situations, although there have been declines in some areas and some species. The best places to find frogs are likely breeding sites, particularly after rain. You will most probably hear frogs as you approach the site, and they can be found by the water's edge, in or on nearby vegetation, or under rocks, logs and other cover.

It is often easier to find frogs at night time, because they are more active then, and their eyes often reflect back a torch beam as two golden spots of light. Triangulation is another useful method for finding a calling frog. Two or more people gather around a calling frog — care must be taken here, as the frog will stop calling if alarmed. As the frog calls, each person points to where they hear the sound. Where these "lines of direction" intersect will usually be where the frog is. The same procedure can be followed at night with torches.

Once the frogs are located, a number of pieces of information need to be recorded, if possible: location, time of day, water temperature, air temperature, weather conditions, calling site and species name. At the very minimum, you need to write down the location, the species name and the date when it was seen.

This information should be recorded in your field record sheet, or in a field notebook. The information built up over time will form a valuable record of the requirements for each species, and their status.

A useful item of equipment, particularly if you are having trouble seeing the frogs, is a portable tape recorder. Notes can be dictated into it for writing into a notebook later, and frog calls can be recorded for comparison with existing identification tapes. If you play its taped call back to a frog, it will usually answer, letting you locate and identify it.

If you can, you should make an estimate of how many individuals are in the area, using a four-point scale which ranges from "rare" to "a few", to "common", to "very large numbers". You may find that the status changes during the year, as many species have set breeding times.

Captive specimens

To learn more about frog behaviour, to observe metamorphosis, or to help identify an unknown species, you may need to keep either tadpoles or frogs in captivity for a short time before releasing them. If you plan to do this, check first to see if it is legal to keep frogs in your State. As this book goes to press there is legislation operating in New South Wales and Victoria, which makes the keeping and handling of eggs, tadpoles and frogs illegal without a permit. Contact your local national parks authority and obtain permission before keeping or handling frogs, or starting a Frog watch program .

Frog eggs

You can easily collect a sample of spawn (eggs) by taking the twigs or bits of vegetation to which the spawn is attached. Foam egg masses can be snipped in half with scissors, and loose eggs can be scooped up in a teaspoon.

Eggs will hatch best if you keep them in the water in which you found them. Tap water contains chemicals which kill the eggs and tadpoles, unless it is left to stand in a clean,open plastic or glass container for a few days.

Be careful to treat the eggs gently when you are transporting them. Too much sloshing on the trip home, and many of them will die. Do not keep them where they will "cook" in hot, direct sunlight.

Tadpoles

You can catch tadpoles in an old vegetable strainer, a tea strainer, a homemade net or an aquarium net (available from most pet shops). They can be stored and transported in plastic buckets or plastic bags but once again, beware of sloshing them around too much.

Once at home, they can be housed in any suitable plastic or glass container (not metal, as it will poison the water). A glass aquarium is perhaps the best from the observation point of view. Do not overcrowd them, and do not overfeed. A cheap aquarium air pump and box filter can help keep the water clean and aerated, but it is not essential. Change about a quarter of the water volume once a week in either case. The container may have a layer of clean sand or aquarium gravel and a few water plants, but this is only necessary to make it attractive. Tadpoles can also be kept in an unfurnished aquarium.

The easiest food to provide is a piece of crushed or lightly boiled lettuce. Only provide small amounts, as too much will rapidly spoil the water. This causes an unpleasant odour and bad conditions for the tadpoles, and requires the water to be changed. As the tadpoles grow, their diet can be supplemented with prepared aquarium fish food. This also should be used sparingly.

As legs develop on the tadpoles, emergent rocks or branches must be provided, or the young frogs (metamorphs) will drown (tadpoles have gills, and breathe in the water; frogs have lungs, and breathe in the air). If the developing frogs seem to have adhesive discs on their toes, you will also need a close-fitting lid, or they will climb out. If you want to raise some frogs, select a small number and release the rest where you found them (they will have a better chance of survival if released at night). Trying to raise too many just produces lots of bony, stunted, starving frogs.

Keeping Frogs

Frogs are easiest to catch at night using a torch. A wet cloth bag, plastic bag or old wet pillowcase will do to transport them. Don't put in too many, as accumulated skin secretions may kill some of the frogs themselves. Also, large frogs will eat small ones.

An aquarium is perhaps the best frog container, as it allows easy observation. A tall or vertical aquarium is better for tree frogs and a conventional aquarium is more useful for terrestrial forms. It need not be watertight, unless you are keeping aquatic species. The base of the aquarium can be covered with clean river sand, damp sphagnum or other moss, or smooth gravel. Do not use garden soil or potting mix, as a rapid build-up of bacteria will occur, resulting in the death of the frog. Burrowing frogs need enough sand to be able to disappear under. Non-burrowing frogs need only about 25 mm of sand.

If plants are to be included, it is best to leave them in small pots, so they can be removed easily for pruning, repair or replacement. The actions of frogs will often flatten more delicate plants. Spiny plants should be avoided for obvious reasons. To provide water, a plastic bowl or cut-down ice-cream container should be sunk into the base material so that the rim of the container is level with the sand. This should be filled with water, and a few pebbles added to allow easy access for the frog.

Some pieces of bark, branch, rock or pieces of terracotta plant pot need to be added next, as most frogs are nocturnal and need to hide during the day. The last requirement is a close-fitting, ventilated lid. The best sort of lid is waterproofed wood, as you can install both nylon mesh ventilation panels and a small door to drop in live insect food. This door is useful in preventing the escape of both frogs and insects whenever food is added. The container should be kept moist — not sodden, and not dry. This can be achieved by spraying the container frequently with water in a cheap plant spray bottle available at most nurseries and supermarkets.

For small species and metamorph frogs, Vinegar Flies (*Drosophila melanogaster*, the standard "fruit fly" of genetics) can be cultured and used. Mosquito larvae added to the water dish will hatch into mosquitoes and be eaten as they emerge. Flies, beetles, spiders, grasshoppers and moths can all be offered to the frogs — the larger the frog, the larger the prey it can swallow. Food must be live and moving, as most frogs refuse dead, immobile prey. Do not use insects that have been sprayed with insecticide. Do not keep frogs together where there is a large size difference — the smaller ones might end up on the menu.

Finally, when you have finished with them, return the frogs to where they were found, as they will have a better chance of survival there. Releasing them on a wet night is best.

Preserving specimens

Sometimes a frog or tadpole will die in captivity, or you will find a not-too-badly damaged specimen on the road. Such specimens should not be wasted, as they can provide the basis for a small reference museum, or they may be sent to a natural history museum to confirm your identification. If time permits, the frog can be arranged in a tray or ice-cream container in a position which shows those features important to identification, e.g. mouth wedged open, one hind leg extended and fingers and toes spread.

Preservative such as alcohol or methylated spirits (65 per cent solution in water) can be added to just cover the specimen, and the container should then be covered and left until the specimen sets. Some researchers make a slit in the belly, or inject preservative to preserve the internal organs. This can usually be omitted, as most frogs have thin, permeable skins.

After that, the specimen can be stored in the same preservative in a water-tight jar. Do not forget to include a label, giving details of place and conditions of collection site, date and collector's name. Weather conditions and temperature are of great value, if you can include them. The species name can be added, if you know it. Only one specimen per jar, unless the collection details are the same for all. A label can be tied to the specimen, provided the ink on the label does not dissolve in alcohol.

The same details should be stored with tadpole specimens, which don't need to be set, just dropped into preservative. Sometimes a specimen will be found that is mortally wounded, but still alive. The best euthanasia for such frogs is to put them into the refrigerator for an hour or two, which puts them into a sleep-like state. Then put them straight into the freezer and leave them overnight. This kills them and freezes them solid. Once they are thawed out, follow the same preservation procedures given above. Once again, you need to check with your local wildlife authorities about keeping preserved frogs, as you will need to obtain a permit in most states.

HOW TO USE THE KEY

Don't panic! The key looks alarming at first glance, but it works. It is there to help you, and can only make your identification more reliable.

This key has been designed for use by students and teachers who may have little experience in handling frogs. All identification features which involve internal examination — even of the inside of the mouth — have been avoided. The features should all be visible to the naked eye, although small specimens will have smaller features, so a magnifying glass may be useful.

The frogs can be hand-held, or observed in a jar while you proceed through the key: some manipulation may be necessary to see the colouring on the back of the thigh or extent of toe webbing. To determine the shape of the pupil, you need to look at frogs in a bright light.

Scientific terminology has been kept to a minimum in both the key and the species descriptions which follow. Those terms included are important and can be easily identified from the illustrations both at the front of the key and within it. The glossary on page 110 will also help to explain the terms used.

The key works by providing a choice of statements. Whichever statement applies to your frog is the correct answer. Many frog species vary quite a lot in colour and appearance, so when you are in doubt, select the choice which is closest to the frog you are identifying. This choice will then lead to a number. This number refers to another series of statements, and so on until you end up with a "Latin" name and a page number. The page number refers to the location in the book of the main description, distribution and illustration of the frog, which you then use to check your identification.

Alternatively, you can look through the pictures in the book and pick the one that looks most like the frog you have. Check the description and distribution to see if this information also fits. As a final check, you can back-track through the key, starting with the key number provided. Each statement number is followed by a number in brackets — this bracketed number refers to the previous statement that led you to this one.

A simplified key below shows the process of identification for Green Leaf Tree Frog, *Litoria phyllochroa.* You can see from this example that the process is simple and logical, although at first sight, the lines and numbers might alarm you a little. Just take it step by step.

1A The frog is small		**2**
1B The frog is large		**3**

2A (1A)	The frog is brown	*Crinia signifera* (p. 10)
2B (1A)	The frog is yellow and black	*Pseudophryne corroboree* (p. 12)
2C (1A)	The frog is green	*Litoria phyllochroa* (p. 14)

You start by choosing either 1A or 1B, which directs you to either group 2, or to group 3. From there, you quickly get a species, and a page number to turn to, where you will find more information, in a species description. Once you have your frog identified, you need to check it by reading the description.

The "Description" for each species covers only those colours and details which will help you identify the frog. Webbing between the toes, for example, is always included, but webbing between the fingers is mentioned only where it is obvious. Only the commoner colour forms are listed, so an albino or other rare colour form may prove difficult to identify. The colours listed are the colours of live frogs — preserved specimens are often different in colour. Metamorphs (juvenile frogs, just after the tadpole stage) are sometimes different from the adult form.

You will notice that tooth details are included under "Description" for each species. If a dead specimen is available, its identity can be narrowed down quite a bit by looking at its dental arrangement. This examination is difficult with a live frog, and may injure the frog, so we advise you not to do it.

The "Calls" section describes the sound the frog makes, in a written form. Although this is of variable use (one person hears "tok", while another hears "thud"), it is better than nothing. This section also provides information on where the males call from, and what time of year they have been heard. This can help in identification and in locating other calling individuals of the same species.

"Habitat" provides details on where the species may be found and where it breeds (if this information is available). In some cases, this will not help in identification, because either not enough is known about the frog, or it has been transplanted from its normal habitat. This can happen when a specimen turns up in a plant nursery, or a specimen is released in one habitat by someone who caught it in another, but such events are rare.

The "Distribution" map indicates by shading where the frog species has been found, or is expected to be found. This may change for some species over a period of time.

"Similar species" lists other frogs which could be mistaken for this species, and shows you how to tell them apart. This won't always be easy, but it should be possible in all the cases listed.

ANATOMY OF A FROG

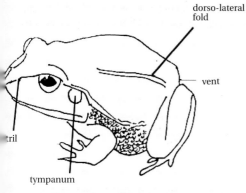

dorso-lateral fold

vent

tril

tympanum

Dorsal surface

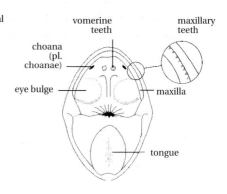

vomerine teeth

maxillary teeth

choana (pl. choanae)

eye bulge

maxilla

tongue

Interior of mouth

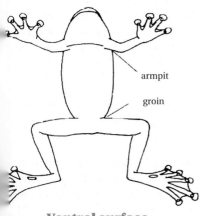

armpit

groin

Ventral surface

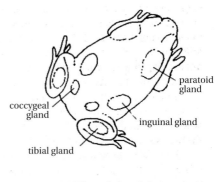

paratoid gland

coccygeal gland

inguinal gland

tibial gland

Gland positions

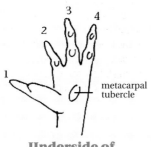

3
2
4
1

metacarpal tubercle

Underside of hand

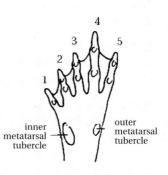

4
3
5
2
1

inner metatarsal tubercle

outer metatarsal tubercle

Underside of foot

THE KEY

See Glossary of terms on page 110

1A Obvious enlarged glands on back of neck (1) **2**
1B Obvious enlarged glands on leg (2) , pupil
horizontal (6)   **59**
1C No obvious enlarged glands on leg or neck **3**

2A (1A) Tympanum hidden (3) . Bright yellow, red
or orange spot in groin and in back of knee.
Small size, 30 mm or less body length   **8**
2B (1A) Tympanum hidden. No red or orange
spot in groin, or at back of knee
Philoria frosti (p. 53)
2C (1A) Tympanum distinct. Grey-brown, olive-
brown or red-brown above. (4) Two bony ridges
running from between nostrils to above eye (4)
Large size: up to 200 mm body length.
Bufo marinus (p. 109)

3A (1C) When constricted, pupil of eye is
vertical
(5)   **4**
3B (1C) When constricted, pupil of eye is
horizontal (6)   **5**
4A (3A) No adhesive discs (8) , and distinct tympanum   **25**
4B (3A) Slightly enlarged tips to fingers and toes,
tympanum indistinct, toes fully webbed
Rheobatrachus silus (p.64)
4C (3A) Frog has large, shovel-shaped
metatarsal tubercle (7) , tympanum indistinct (9) ,
toes fully webbed   **60**

5A (3B) Frog has adhesive discs on fingers and
toes (10), or there is a notch on dorsal surface of
toe before the last joint.   **6**
5B (3B) No discs or notch present (8)   **7**

6A (5A) Toes webbed to some extent   **24**
6B (5A) Toes not webbed, but may be fringed
(11) *Taudactylus diurnus* (p. 65)

7A (5B) Toes webbed to some extent, even if
only a trace, and tympanum indistinct, unless
there are pink stripes on body   **54**
7B (5B) Toes not webbed, and tympanum either
hidden or indistinct   **11**
7C (5B) Toes webbed, and tympanum distinct
29

8A (2A) Both parotoid and inguinal glands greatly
enlarged (12)   **9**

8B (2A) Inguinal glands not enlarged (13)   **10**

9A (8A) Scarlet patches in groin and behind knees, and belly whitish with brown flecking. No obvious dark patch on snout
Uperoleia capitulata (p. 67)
9B (8A) Yellow or orange patches in groin and behind knee, belly is a grey colour. Often there is a dark triangular patch on snout
Uperoleia rugosa (p. 71)

10A (8B) Body patterned above, with dark brown or blackish spots, bars and reticulations, usually with a pale triangular patch on snout. Belly is pale grey, speckled with dark purplish-brown. This grey and brown colouring is patchy — belly never uniformly pigmented
Uperoleia laevigata (p. 70)
10B (8B) Dorsal surface dark brown, with little patterning. Parotoid glands enlarged. Belly totally uniform dark grey to blue-black
Uperoleia tyleri (p. 72)
10C (8B) Dark brown above, with little patterning. Parotoid glands barely raised above surface of head and neck. Belly cream, with dark grey or brown flecking, and uniformly pigmented
Uperoleia fusca (p. 69)
10D (8B) Back grey, with yellow and dark brown patterning. Parotoid glands enlarged. Belly brown with white flecking, and uniformly pigmented   *Uperoleia martini* (p. 70)

11A (7B) Toes not fringed (15), and belly smooth or slightly granular (14)   **12**
11B (7B) Toes long and fringed (17), and belly granular (16). Small, with bodies less than 35 mm   **14**
11C (7B) Toes fringed, belly smooth and mottled dark brown and white: pink, white and dark brown mottling on lower belly and hind limbs
Crinia tasmaniensis (p. 25)

12A (11A) Belly not boldly marbled   **13**
12B (11A) Belly boldly marbled (18) in black and white or yellow   **20**

13A (12A) Belly cream-coloured, with heavy brown mottling. Fingers and toes have slightly swollen tips (39)   *Assa darlingtoni* (p. 23)

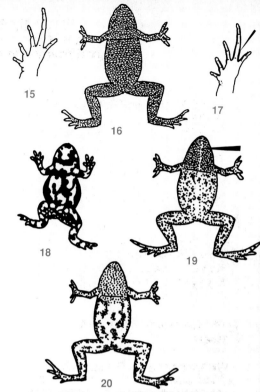

13B (12A) Belly white, or light grey with grey or brown flecks all over. In adult males, throat is yellow   *Geocrinia laevis* (p. 30) or
Geocrinia victoriana (p. 32)
13C (12A) Belly white or bright yellow, and throat only is flecked with brown   **53**

14A (11B) Line of white or pale dots running down the throat (19)   **15**
14B (11B) No line of white or pale dots running down throat (20)   **16**

15A (14A) Belly white or light brown. Very little mottling or flecking   *Crinia tinnula* (p. 25)
15B (14A) Belly boldly marked with black and white marbling (mainly southern specimens)
Crinia signifera (p.25)

16A (14B) Belly brown with pale spots. Base of each arm, the groin, and back of thighs are all bright orange-red   *Paracrinia haswelli* (p. 51)
16B (14B) Belly not brown with pale spots. No orange-red patches as in 16A above   **17**

17A (16B) Palm of hand has tubercles (21)   **18**
17B (16B) Palm of hand is smooth (22)   **19**

18A (17A) Belly has black and white mottling
Crinia signifera (p. 25)
18B (17A) Belly dirty white, lightly flecked with black *Crinia sloanei* (p. 25)
18C (17A) Belly pale and unspotted, pale muddy brown above, with little patterning. Adult' body less than 18 mm long
Crinia deserticola (p. 25)

19A (17B) Belly has black and white mottling
Crinia riparia (p. 25)
19B (17B) Belly grey, sometimes with darker flecks, and palms are smooth
Crinia parinsignifera (p. 25)

20A (12B) Tip of the fourth toe extends to tip of snout or beyond (23) **21**
20B (12B) Tip of the fourth toe extends only to neck or eye (24) *Pseudophryne dendyi* (p. 60)

21A (20A) Belly mottled black (or brown), and yellow **22**
21B (20A) Belly mottled black and white **23**
21C (20A) Belly mottled black and white with throat and underside of limbs orange or flesh coloured
Pseudophryne semimarmorata (p. 63)

22A (21A) Back brown, with irregular darker marbling *Pseudophryne major* (p. 61)
22B (21A) Back not brown, with irregular darker marbling **23**

23A (21B, 22B) Frog dark above, with red triangle on forehead and variegated red markings on back *Pseudophryne australis* (p. 57)
23B (21B, 22B) Frog dark above, uniform or with reddish variegations
Pseudophryne bibroni (p. 57)
23C (21B, 22B) Bright or brick red on head and back. Sides purplish-black, with the two colours strongly contrasted laterally
Pseudophryne coriacea (p. 59)
23D (21B, 22B) A striped frog, with brilliant yellow and shiny black markings
Pseudophryne corroboree (p. 59)

24A (6A) Tympanum distinct **32**
24B (6A) Tympanum indistinct or hidden **52**

25A (4A) Hind limbs have dark cross-bars **26**

25B (4A) Hind limbs have no dark cross-bars. Skin of rough appearance, usually grey. Males possess black spines on backs of hands. Tympanum distinct, and toes webbed at the base only. There is a large, pale, shovel-shaped inner metatarsal tubercle
Heleioporus australiacus (p. 33)

26A (25A) Toes at least two-thirds webbed **27**
26B (25A) Toes webbed at the base only
Lechriodus fletcheri (p. 34)

27A (26A) Upper half of eye has a pale blue curve under the eyelid, and belly is white or pale yellow. Usually, no dark spots or blotches on the sides *Mixophyes balbus* (p. 44)
27B (26A) No blue colour in upper half of the eye, unless the belly is yellowish. Dark spots or blotches on sides **28**

28A (27A) Broad, dark cross-bars on hind limbs, with intervening lighter areas which join together on backs of thighs to form uniform dark background with scattered pale spots (25). Upper half of eye pale gold *Mixophyes iteratus* (p.44)
28B (27A) Dark cross-bars on hind limbs narrow and distinct, broadening on front and hind edges to form triangular patches which can be seen from beneath (26). Upper half of eye is dark. Upper lip pale creamy white without dark blotches *Mixophyes fasciolatus* (p. 44)
28C (27A) Cross-bars on hind limbs are as in 28B (*Mixophyes fasciolatus*). Upper half of eye is silvery or pale blue, and upper lip is brownish with one or more dark brown blotches
Mixophyes fleayi (p. 44)

29A (7C) Toes less than half-webbed **30**
29B (7C) Toes half-webbed, or more than half-webbed **31**

30A (29A) Toes nearly half-webbed. Backs of thighs are uniform greenish or greyish, and there is a dorso-lateral fold on each side of back (27)
Cyclorana novaehollandiae (p. 75)

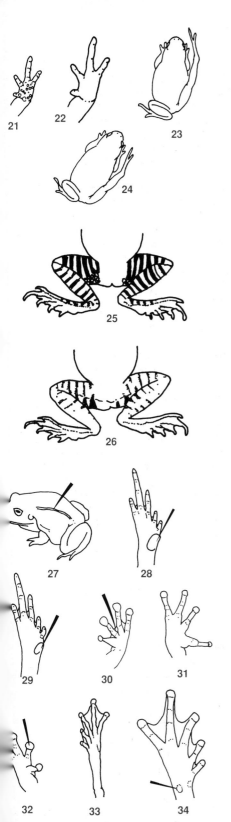

30B (29A) Toes one-third webbed. Metatarsal tubercle is at least as long as its distance from the tip of innermost toe (**28**). Colour pattern indistinct. Slightly to extremely warty, with skin folds on dorsal surface. Back of thighs grey-brown, spotted or mottled with white
Cyclorana verrucosa (p. 76)
30C (29A) Toes one-quarter webbed or less, and body dark brown above, with distinct pale grey-brown blotches. Backs of thighs dark brown with a few pale flecks, and the length of the metatarsal tubercle is shorter than the distance from it to tip of innermost toe (**29**)
Cyclorana brevipes (p. 74)

31A (29B) Toes entirely webbed, and backs of thighs grey with darker flecks
Cyclorana platycephala (p.75)
31B (29B) Toes half-webbed. Backs of thighs dark brown or black, with accentuated white spots   *Litoria alboguttata* (p. 78)

32A (24A) Fingers at least one-third webbed (**30**)
33
32B (24A) Fingers either not webbed at all, or less than one-third webbed (**31**)   **37**

33A (32A) Body mostly green above   **34**
33B (32A) Body not mostly green, but some green patches or tiny green flecks   **35**
33C (32A) Body has absolutely no trace of green colour   **36**

34A (33A) Body is a uniform green above, often with a few white spots laterally. Back smooth, with very tiny pores. No dark stripe through eye and tympanum   *Litoria caerulea* (p.83)
34B (33A) Uniform green above, never white-spotted. Back finely granular. Tympanum covered with granular skin (**38**). No dark stripe through eye and tympanum
Litoria gracilenta (p.91
34C (33A) Similar to 34B *(Litoria gracilenta)*, but larger. Distinct tympanum covered with smooth skin (**37**). No dark stripe through eye and tympanum.
Litoria chloris (p. 85)
34D (33A) Light green above, often with fawn spots. A dark stripe runs from nostril through eye, and over tympanum
Litoria burrowsae (green form) (p. 81)

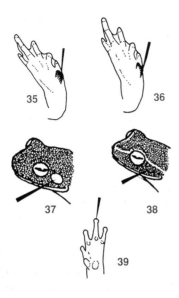

35

36

37

38

39

38A (37A) Fingers have a rudiment of web, and toes are approximately two-thirds webbed. Brown above, with a few faint spots on groin, and no green on the body at all. Backs of thighs, groin and armpits are all orange or red, with no black markings   *Litoria jervisiensis* (p. 93)
38B (37A) Fingers with rudiment of web. Toes approximately two- thirds webbed. Cream brown to reddish brown above, with one to three black spots in groin. Backs of thighs orange with some black markings

Litoria revelata (p. 102)
38C (37A) Fingers have basal web at most; toes are at least half-webbed. Backs of thighs are red, orange or brown, with no marbling. Armpits and groin are not bright orange or red, unless the body is green or brown   **39**

39A (38C) Body is uniform green above   **40**
39B (38C) Body has some green colouration only   **41**
39C (38C) Body has no green colouration at all

Litoria rubella (p. 104)

35A (33B) Variegated brownish above, with scattered emerald flecks. Backs of thighs have black and yellow marbling. A large inner metatarsal tubercle, and iris is silvery   *Litoria peronii* (p. 98)
35B (33B) Variegated brownish above, with scattered emerald flecks. Backs of thighs have yellow and brown marbling. Small oval inner metatarsal tubercle, and iris is golden   *Litoria tyleri* (p. 107)
35C (33B) Dark brown, with light green and brown patches and flecks. A dark stripe runs through eye and over tympanum. Back of thigh pale brown

Litoria burrowsae (brown form) (p. 81)

36A (33C) Brown or bluish-grey above, with lighter zone each side of back. Backs of thighs usually have a yellowish tinge   *Litoria dentata* (p. 87)
36B (33C) Variegated brownish above. Backs of thighs yellow with black bars. Top half of iris is rust-coloured, and lower half silvery grey   *Litoria rothii* (p. 103)

37A (32B) Toe discs are obviously wider than toes (32) **38**
37B (32B) Toe discs are small: same width, or only a little wider than toes (33)   **42**

40A (39A) Fingers have distinct basal web. Tympanum is green   *Litoria phyllochroa* (p. 99)
40B (39A) Fingers have distinct basal web. Tympanum is brown   *Litoria pearsoniana* (p. 97)

41A (39B) Fingers have just a trace of web. Body brownish or greyish above, sides green and underside of legs bright red   *Litoria citropa* (p. 86)
41B (39B) Fingers have distinct basal web. Body marbled dark slate and dull green above, with dark marbling and flecking   *Litoria piperata* (p. 100)

42A (37B) Backs of thighs not spotted unless they are bluish, when yellow or cream spots may be present. No distinct outer metatarsal tubercle   **43**
42B (37B) Backs of thighs spotted or marbled. A small outer metatarsal tubercle (34)   **48**

43A (42A) Backs of thighs not bluish or spotted   **44**
43B (42A) Backs of thighs blue, with or without spots

47

44A (43A) A broad, brown band on back   **45**
44B (43A) No broad brown band on the back   **46**

45A (44A) Snout rounded. Fingers unwebbed, and toes half-webbed. Body brown above, with darker reticulations, sometimes with green patches. Backs of thighs yellowish to bright reddish-orange in mountain populations. Yellow to orange in the groin, with black spots or marbling. Calls all year round   *Litoria verreauxii* (p. 107)
45B (44A) Snout rounded, with conspicuous broad, dark brown band from between eyes to anus. Backs of thighs yellow to orange. No black spots or marbling in groin
    Litoria ewingii and *Litoria paraewingi* (p. 89 & 97)

46A (44B) Snout rounded. Fingers have a rudiment of web, and toes half to three-quarters webbed. Back either all green, all brown, or frog has brown legs with green back. A white stripe from angle of jaw below eye to shoulder. Backs of thighs orange. Discs small but distinct, and tympanum is brown   *Litoria fallax* (p. 90)
46B (44B) Snout pointed. Body light brown to light green above, fingers webbed at base, and backs of thighs orange. A white stripe from angle of jaw to base of hind leg (it may sometimes break up into a series of white blotches), and tympanum brown
    Litoria olongburensis (p. 96)
46C (44B) Snout rounded. Body green, with brown spots on back. Backs of thighs purplish brown, and tympanum green   *Litoria cooloolensis* (p. 87)

47A (43B) Snout pointed, and fingers unwebbed. Blue-green above, usually with coppery-gold stripes and spots. Backs of thighs dark blue. Skin smooth
    Litoria aurea (p. 79)
47B (43B) Snout pointed. Colour as above (*Litoria aurea*), and skin covered in warts, tubercles and short skin folds. Backs of thighs bluish   *Litoria raniformis* (p.101)
47C (43B) Snout pointed. Like 47B (*Litoria raniformis*), but with large yellow or cream spots in groin and on back of thighs   *Litoria castanea* (p. 84)

48A (42B) Backs of thighs dark brown, with lighter spots or marbling   **49**
48B (42B) Backs of thighs not dark brown with lighter spots   **50**

49A (48A) Snout pointed, back warty, brownish variegated with lighter colouring. Backs of thighs dark brown, with a few irregular lighter spots   *Litoria freycineti* (p. 90)
49B (48A) Snout pointed, back smooth, dull grey with irregular darker marbling and flecking. Backs of thighs dark brown with pale spots   *Litoria booroolongensis* (p. 80)
49C (48A) Snout pointed, back warty. Backs of thighs are mottled dark brown and white or yellow
    Litoria inermis (p.92)

50A (48B) Snout very pointed. Back has longitudinal folds of skin, and irregular alternating bands of dark and light brown. Fingers without webbing. Backs of thighs yellow with brown lines and spots   *Litoria nasuta* (p. 95)
50B (48B) Back smooth with no skin folds; colouring different from 50A   **51**

51A (50B) Body either uniform brown, or peppered with darker colouring. Backs of thighs marbled black and white, and groin black and yellow. Dark eye stripe does not cover tympanum   *Litoria lesueuri* (p. 95)
51B (50B) Snout pointed. Body either uniform bluish-grey or greyish-brown, or with faint darker variegations. Backs of thighs are marbled black and yellowish-white, groin is not black and yellow. Dark eye stripe almost completely covers tympanum   *Litoria latopalmata* (p. 94)
51C (50B) Snout rounded. Body rich red-brown or chocolate above, and groin and backs of thighs brilliant blue, blue-green or green, with or without black blotches. Dark eye stripe almost completely covers tympanum
    Litoria brevipalmata (p. 81)

52A (24B) Grey or olive-green above, with darker mottling or marbling. Skin rough (shagreened) with some low warts, and toes fully webbed   *Litoria spenceri* (p. 105)

52B (24B) Green or olive above. A broad golden stripe, underlined by a broad black stripe, running from snout through eye and down the side, breaking into patches towards the back. The skin smooth or finely granular above, and toes two-thirds webbed. Found in NSW
Litoria subglandulosa (p. 105)
52C (24B) Light to dark green above. A broad golden stripe underlined by a narrow black stripe, the two colours becoming mottled along flanks. Toes three-quarters webbed. Found in Victoria
Litoria phyllochroa (southern form) (p. 199)

53A (13C) Alpine frogs, usually found in moss and damp soil in wet sclerophyll forest, rainforest and Antarctic Beech. Inner metatarsal tubercle is small and blunt, and first finger is much shorter than second. A brown band from eye to arm, and another along side of body. Toes free of webbing, and belly white
Philoria sphagnicolus (p. 55)
53B (13C) As above (53A), including white belly, but without dark band on side.   *Philoria loveridgei* (p. 54)
53C (13C) Body variable, from yellow to claret red, to black above, with or without V-shaped bands on back. Belly bright yellow   *Philoria kundagungan* (p. 54)

54A (7A) Rotund globular frog   **55**
54B (7A) Not rotund globular frog; webbing present, but reduced, and toes may be fringed   **56**

55A (54A) Toes have one-quarter webbing. A large, pale, shovel-shaped inner metatarsal tubercle. Back patterned in light and dark brown   *Limnodynastes ornatus* (p. 39)
55B (54A) Toes slightly webbed. A large, pale, shovel-shaped inner metatarsal tubercle. Body bright yellow or green, often with cross-shaped pattern in black, red and green warts on the back   *Notaden bennettii* (p. 50)

56A (54B) Toes have just a trace of web. Belly either black with white spots, or bold black and white marbling. A red patch in groin and on hind edge of hind leg
Adelotus brevis (p. 21)

56B (54B) Toes webbed to some degree. Belly white **57**

57A (56B) Pink or red stripes dorso-laterally on back
Limnodynastes salmini (p. 41)
57B (56B) No pink or red stripes on back   **58**

58A (57B) First finger a little shorter than second. Back has bands and stripes along it, sometimes with a pale stripe. Snout pointed and prominent. Toes almost free of webbing   *Limnodynastes peronii* (p. 40)
58B (57B) First finger a little shorter than second. Back pale brown or grey, with darker irregular spots. Snout not prominent. Toes have distinct basal webbing
Limnodynastes fletcheri (p. 37)
58C (57B) First finger a little shorter than, or equal to, the second. Body olive above, with darker rounded blotches; often there is a yellow stripe down the centre. Toes have basal webbing and slight fringe
Limnodynastes tasmaniensis (p. 42)

59A (1B) Toes one-quarter or less webbed. Belly white or pale yellow, and may have black or grey mottling. No scarlet in groin   *Limnodynastes dumerilii* (p. 36)
59B (1B) Toes one-third or half-webbed. Belly yellow, usually without mottling. No scarlet in groin
Limnodynastes interioris (p. 38)
59C (1B) Toes have a basal webbing, and belly white or yellow. Groin and backs of thighs have scarlet markings
Limnodynastes terraereginae (p. 43)

60A (4C) Large black inner metatarsal tubercle (**35**). Skin has numerous small warts on the back, and there is loose groin skin from body to knee
Neobatrachus sudelli (p. 49)
60B (4C) Large black inner metatarsal tubercle. Skin has numerous small warts above, and groin skin not loose, or extending from body to knee
Neobatrachus pictus (p. 48)
60C (4C) Inner metatarsal tubercle black or brown edged only (**36**). Skin smooth above; groin skin extending from body to knee   *Neobatrachus centralis* (p. 47)

FAMILY: MYOBATRACHIDAE

These are sometimes collectively known as the Southern Frogs, and they are confined to the Australian region. Generally, the frogs in this family lack the adhesive toe discs found in the Hylidae family, although some genera (such as *Assa* and *Taudactylus*) have expanded toe tips. These toe tips can sometimes cause confusion in identification.

The family is found all over Australia and Tasmania, with twenty genera recognised by Cogger (1992), although there are only sixteen genera in the area covered by this book.

Within these sixteen genera, there is a wide range of lifestyles, habits, and body forms.

Myobatrachidae means "muscle" frog family, after *Myobatrachus*, the first genus named/described in this family.

GENUS

Adelotus

This genus is restricted to eastern Queensland and New South Wales. It contains only the one species, Adelotus brevis. *The most distinctive feature of the genus is the two large tusks ("dentary pseudo-teeth") in the lower jaw of the male. An unusual feature is that the male grows larger than the female, a reversal of the usual situation among frogs.* Adelotus *means "unseen".*

Tusked Frog
Adelotus brevis

brevis means "short"

Key No. 56A
Description: The males and females of this species look quite different from each other. The males can reach 50mm, while the females are about 40mm. The males also have a larger head, a different belly pattern and much larger "tusks" in the lower jaw. The body colouring above is similar in both sexes, ranging from olive green to almost black above, with an irregular variegated pattern.

Adelotus brevis

Adelotus brevis Female   40mm

There is usually a butterfly-shaped marking between the eyes. The limbs are usually barred or banded with darker markings. The belly of the male is black with white spots, while the female has a vividly black and white marbled belly. Both sexes have black, and red or orange marbling in the groin and on the back edge of the hind leg. The skin is rough, with low ridges and warts. The toes have only a trace of webbing, and the fingers have no web. The vomerine teeth are in two small clusters behind the choanae, and there is a pair of enlarged teeth in the lower jaw.

Call: A single "tok" or "cluck" repeated several times a minute, similar to the call of *Limnodynastes peronii*. It is usually difficult to locate: the males call from the water, hidden among vegetation or behind logs and rocks in the water, throughout the year.

Habitat: This species is usually found associated with water in rainforest, wet sclerophyll forest, or grassland which is sometimes flooded. It is most frequently found under cover beside puddles, streams and ditches.

Similar species: The larger size distinguishes it from *Uperoleia* spp., as will the butterfly-shaped mark between the eyes, which *Uperoleia* lacks. If you are still in doubt, a careful and gentle examination of the lower jaw will reveal the tusks. The bright red patches in the groin and the back of the thigh should distinguish it from *Pseudophryne* and *Crinia* species. The lack of bright colour in the armpit and different belly colour should distinguish it from *Geocrinia* and *Paracrinia*.

GENUS

Assa

*This genus is restricted to the coastal ranges
around the New South Wales/Queensland border.
There is one species in the genus. The most distinctive feature of the genus is the twin pouches, one on
each side, in the male.
Assa means "dry nurse".*

Pouched Frog
Assa darlingtoni

darlingtoni is after P.J. Darlington

Assa darlingtoni

Key No. 13A
Description: This tiny frog is usually
20-30mm in length. It ranges from
grey to red brown above, with variable darker markings frequently in
the form of two inverted "V"s, one
starting between the eyes with the
arms of the V extending towards the
shoulders, the second starting mid-
back, with the V arms extending into
the groin. In other specimens this is
replaced by an irregular band of
darker colour running from between
the eyes to the groin. There may be a

Assa darlingtoni   30mm

dark broken stripe from the nostril through the eye and down each side. The sides are usually dark grey to black. The belly is cream or white, with a brown mottled throat. The fingers and toes are unwebbed and unfringed, and the first finger and toe are rudimentary, but all digits have slightly swollen tips. The skin is smooth on the back and the belly, but the sides may be rough or warty. There are no vomerine teeth.

Call: A series of rapidly repeated "Eh..Eh..Eh..Eh..Eh..Eh". Usually between six and ten notes. Males call from leaf litter, rocks or logs.

Habitat: It is found in Antarctic Beech forest and rainforest in mountainous areas. These frogs do not need free water for breeding, as eggs are laid on the ground. On hatching, the tadpoles enter pockets on the males' hips, where they metamorphose to emerge as fully formed froglets. They spend most of the time in damp leaf litter, or under rocks and rotten logs.

Similar species: *Philoria loveridgei* is similar and found in the same area, but has well-developed fingers and toes without swollen tips, and it lacks any distinctive pattern on its back. *Philoria kundagungan* has a bright yellow belly, and *Taudactylus diurnus* has fringed toes with obvious discs.

GENUS

Crinia

The Crinia *frogs are all small frogs. They may have fringed toes, but the toes are never webbed. Species of* Crinia *are found in every state of Australia. The membership of the genus has had several recent changes. Some were once thought to be a single species, but now have been split up into several species. Some members have even been transferred to other genera (*Paracrinia, Ranidella*), and then moved back again, in the case of* Ranidella. *To make matters even more confusing, most members of the* Crinia *genus are very similar in appearance, so that they can often be told apart only by their calls or by genetic analysis. The species in the area covered by this guide should all be distinguishable by appearance. The meaning of* Crinia *is obscure. It possibly means "waterlily pad".*

Crinia
deserticola
18mm

Brown Froglets
Crinia deserticola *(18C)*
Crinia parinsignifera *(19B)*
Crinia riparia *(19A)*
Crinia signifera *(15B, 18A)*
Crinia sloanei *(18B)*
Crinia tasmaniensis *(11C)*
Crinia tinnula *(15A)*

desterticola means "desert dwelling"; the meaning of *parinsignifera* is obscure but *signifera* means "sign bearer"; *riparia* means "riverbank"; *sloanei* is from J.F. Sloane, from whose property specimens were collected; *tasmaniensis* means "from Tasmania"; *tinnula* means "tinkling".

Key Nos. 11C, 15A, 15B, 18A, 18B, 18C, 19A, 19B

Descriptions: As all the above frogs are extremely similar in appearance and behaviour (as far as is known), they are treated together here, with the differences between them listed

Crinia deserticola

Crinia parinsignifera

Crinia riparia

25

Crinia signifera

in the "Call" and "Similar species" sections.

The colours range from light grey through brown to almost black. The patterning can be light-coloured on the back with black sides; or grey or brown with irregular darker patches; or a dark vertebral band bordered by a light brown or grey band with dark bands below that; or a dark vertebral

Crinia signifera
30mm

Shown here are some of the distinctive variations which may occur within a single population of this widespread species

band between skin folds and other dark lines following crests of other skin folds. The sides are black. There is great variability, even within a single batch of young. In size, these frogs range from up to 20mm to 30mm. The belly colours vary according to species (see "Similar species"), and skin may be smooth, have irregular small warts, or have

Crinia sloanei

G.LITTLE

Crinia parinsignifera   20mm

raised longitudinal skin folds. Bellies are granular. The fingers and toes are unwebbed, but the toes are fringed. There are no vomerine teeth.

Calls: *C. deserticola* has a melodious, sparrow-like chirping call. There are no available details on the time of year that it calls.

C. parinsignifera has a drawn-out "low squelch" sounding call. It calls from water's edge, not from water itself, all year round.

C. riparia has a long, harsh "cra-a-a-ak" sound. It calls from the water's edge during spring.

C. signifera sounds like "crick-crick-crick". It calls either from the water's edge, or while floating among vegetation, throughout the year.

C. sloanei has a short, peeping call: "chick..chick". It calls while floating among vegetation, throughout the year.

C. tasmaniensis has a rapid "ek .. ek . . ek .. ek" noise, almost a bleating sound. The males call in spring, at or

Crinia tasmaniensis

near the edge of the water.

C. tinnula has a short high-pitched ring: "tching..tching". Apparently a late winter breeder, it calls from May to September.

Habitats: *C. deserticola* is found in damp areas (creek beds, clay pans, etc) associated with broad river channels in semi-arid regions.

C. parinsignifera is usually found in areas of woodland which are covered with water, open areas, and disturbed sites.

C. riparia is usually found beside fast-flowing streams and creeks.

Crinia tasmaniensis   30mm

Crinia riparia   25mm

Crinia sloanei   25mm

C. signifera is found in almost all habitats from the mountains to the coast, from wet sclerophyll forest through grassland to disturbed areas, usually associated with water, and it is sometimes found in suburban fishponds.

C. sloanei is found in woodland, grassland and open or disturbed areas, usually associated with inundated areas.

C. tasmaniensis is believed to be largely aquatic, and so is usually associated with ponds, creeks and streams in areas above 600 m.

C. tinnula is believed to be confined to acid paperbark swamps in the wallum country.

Similar species: *Crinia* can be distinguished from *Uperoleia* by the lack of parotoid glands; from *Paracrinia* and *Assa* by its very granular belly; and from *Pseudophryne* by a far more subdued mottling on the belly. Within the genus, the distinguishing features are as follows:

C. deserticola has a pale and unspotted belly, and is pale muddy brown above, with little patterning. The adults are usually less than 18mm long. The palm has tubercles.

C. parinsignifera has a grey belly, sometimes with darker flecks, and it

Crinia tinnula

29

Crinia tinnula
30mm

G.A.HOYE

has a smooth palm.
C. riparia has a black and white mottled belly and a smooth palm. It has only been recorded around the Flinders Ranges, so far.
Crinia signifera has a belly blotched in black and white. The throat and

chest of males are dark grey, usually with no mid-line of pale spots. The palm has tubercles.
C. sloanei has a belly which is dirty white, lightly flecked with black. The palm has tubercles.
C. tasmaniensis has a brown, white and pink marbled belly and a tubercular palm. It often has a red-brown or orange stripe running from nostril to groin. It has only been recorded from Tasmania, so far.
C. tinnula has a white or light brown belly with a little mottling or flecking, and a mid-line of white dots on the throat. The palm has tubercles.

GENUS

Geocrinia

The members of this genus are all restricted to the south, with none known from Queensland or the Northern Territory. The two species in this guide are extremely similar, but can be distinguished by the calls of the males, and by their distributions.
While they are superficially similar to Crinia, *the frogs of this genus lack the fringed toes found in the south–east –Australian species of* Crinia.
Geocrinia *means "earth-" crinia.*

Southern Smooth Froglet
Geocrinia laevis

laevis means "smooth"

Key No. 13B
Description: This frog is up to 35mm long. It is grey or brown above, with black-edged red spots and some darker blotches. The belly

is white or grey with dark brown or grey flecks and mottling. The undersides of the arms and legs are mottled with pink and there is a pink patch in each groin and armpit, usually mottled or outlined in black. The upper skin is smooth with scattered low warts, and the belly is smooth. The toes are unwebbed and

Geocrinia laevis

Geocrinia laevis
35mm

Geocrinia victoriana   35mm

M.G.SWAN

Geocrinia
victoriana

unfringed. There are vomerine teeth present, but these are not obvious, and the frog also has maxillary teeth. **Call:** A "cra-a-a-k .. cra-a-a-ak .. cra-a-a-k" sound. The males call from the ground in cover such as leaf litter and grass tussocks, in late summer.

Habitat: It is found in sclerophyll forest, shrubland, farmland and disturbed areas. Usually heard and found around areas which get flooded. Eggs are laid in clumps on land, and hatch when the area is flooded by rain.

Similar species: This species is most similar to *Geocrinia victoriana*, from which it can readily be distinguished in the field by call and distribution. It can be distinguished from most *Crinia* by the presence of pink in the groin and under the limbs, and from *Crinia tasmaniensis* by its lack of fringed toes. *Uperoleia* species either have parotoid glands or enlarged inguinal glands, which are missing in this species. *Paracrinia* has a brown and white spotted belly, without the pink.

Victorian Smooth Froglet
Geocrinia victoriana
victoriana means "Victorian"

Key No. 13B
Description: A smallish frog, up to 31mm long. It is grey or brown above, with black-ringed red spots and some darker flecks and blotches. The belly is white or pale grey with brown or grey flecking. The underside of the arms and legs have pink blotches or patches. The armpit and

groin are also pink with black markings. The skin is smooth above, with a few scattered low warts, and the belly is smooth. The toes are unwebbed and without fringes, and the inner finger and toe are much smaller than the others. There are vomerine teeth present, although they are not obvious, and the frog has maxillary teeth.

Call: Sounds something like "craaak.. craaak.. crik.. crik.. crik.. crik.. crik...". Males call from the ground near the breeding site and have been heard calling throughout most of the year.

Habitat: It can be found in damp situations in most kinds of habitat. It is often found beneath rocks, logs, leaf litter, and old housing debris and rubbish near areas which are flooded by rain.

Similar species: This species is easily distinguished from members of the genus *Crinia* by the pink patches in armpits, groin and underside of limbs. It can be distinguished from members of the genus *Uperoleia* by the lack of enlarged glands on the back. It can be distinguished from *Paracrinia haswelli* by the grey or white, smooth belly (light brown in *Paracrinia*) and the pink ventral patches. It can be distinguished from *Geocrinia laevis* by its call and distribution.

GENUS

Heleioporus

Apart from the single species listed here, this genus is found only in south-western Australia. It is large, plump and warty, and it is sometimes mistaken for the introduced cane toad. The differences are easy to spot: Heleioporus *has a vertical pupil in its eye, and does not have enlarged parotoid glands like the cane toad.The males of this genus often have enlarged black spines on the dorsal surfaces of their hands, and they call from burrows.*
Heleioporus *means "marsh hole-dweller"*

Heleioporous australiacus   95mm

Giant Burrowing Frog
Heleioporus australiacus
australiacus means "southern".

Key No. 25B
Description: This is a large species, reaching 90mm in length. It is usually grey, dark chocolate brown, or black on the back, and white on the

33

belly. There are usually a few yellow spots on the sides, and a yellow stripe on the upper lip from the hind edge of the eye to the hind edge of the tympanum. The skin is rough and warty, and breeding males often have pronounced black spines on the backs of fingers. The belly is granular. The toes have rudiments of webbing and there is a divided flap in the front corner of the eye. There are prominent vomerine teeth between the choanae.

Calls: An owl-like call "oo..oo..oo". Males call either from burrows or in the open. They have been known to call while buried. The species breeds during summer and autumn after rain.

Habitat: It is mostly restricted to the Hawkesbury Sandstone. Usually found around sandy creek banks, with crayfish burrows in this area.

Similar species: *Heleioporus australiacus* can be distinguished from *Limnodynastes dumerilii*, *Limnodynastes terraereginae* and *Limnodynastes interioris* by its vertical pupil and lack of a tibial gland. It can be distinguished from *Neobatrachus* by the pale metatarsal tubercle and from *Notaden* by its colour.

Heleioporous australiacus

GENUS

Lechriodus

Only one species of this genus is known in Australia, although other species are known in Papua New Guinea. The Australian species is found only in eastern Queensland and New South Wales. The frog often has a rough sandpapery feel, and it usually has banded limbs.
Lechirodus *means "slanting tooth".*

Fletcher's Frog
Lechriodus fletcheri

fletcheri is after H. Fletcher, the Australian zoologist.

Key No. 26B
Description: This frog is up to 50mm long. It is reddish-brown to nearly black above (although it is often a fawn colour), with occasional

Lechriodus fletcheri. Male and female in amplexus   50mm

darker markings, usually with a transverse darker bar between the eyes. The arms and legs have bars of dark grey-brown. The sides of the head are usually darker brown, but the tympanum is the same colour as the back. There is a dark stripe from the nostril through the eye to the shoulder. The belly is white and the skin is granular to smooth above and smooth on the belly. Males in breeding condition have a rough skin.

Lechriodus fletcheri

There are several skin folds, one from the eye to the shoulder, an X-shaped fold between the shoulders, and usually some on the lower back and legs. The toes have a trace of web, as do the fingers. The vomerine teeth are prominent as long straight ridges behind the choanae.

Calls: The call lasts about one second. "Gar.r..r.up". Males call while floating in the water. A summer breeder, it is rarely seen except during summer rains.

Habitat: This is a rainforest and wet sclerophyll forest inhabitant, usually found in or close to the breeding pools and streams. The tadpoles are highly cannibalistic.

Similar species: The frog is similar to *Mixophyes*, from which it can be distinguished by its reduced webbing, horizontal pupil, and the rough, sandpapery texture of its skin.

GENUS

Limnodynastes

This genus has members in all states,
and one species (L. tasmaniensis) is found in
every state (but not in the Northern Territory).
Many of the species seem to benefit from human
works, being found in farm dams, ditches, even
suburban fish ponds. All species make a frothy egg
mass, and the females in breeding condition often
have flanges on their fingers to help them mix air
into the spawn. The genus includes both burrowers
from arid areas and pond and creek dwellers which
are usually associated with water. The pupil is usu-
ally horizontal, although there may be a ventral
extension, making the pupil look like an inverted
teardrop. There is usually a glandular
stripe below the eye.
Limnodynastes *means "lord of the marshes".*

Banjo Frog, Eastern Pobblebonk

Limnodynastes dumerilii

dumerilii is after A. Dumeril, the French zoologist.

Key No. 59A
Description: A variable species, up to 70mm long, ranging from pale grey to olive green, dark brown or black. Some specimens may have a pale-coloured stripe down the back. There are usually irregular darker markings on the back. A pale raised stripe runs from below the eye to above the arm. Above this, a dark stripe runs from the eye to the tympanum. It is white or yellow in the groin and white below, sometimes mottled with dark grey. The skin is smooth to warty on the back, and smooth on the belly. The toes are a trace to a quarter webbed. There are prominent vomerine teeth behind the choanae.

Calls: A single banjo-like "plonk" or "bonk" repeated at intervals. One frog calling will usually trigger several nearby frogs to call in rapid succession, hence its common name. Males usually call while almost submerged, often from burrows in the bank or under overhanging vegetation, etc. Calls may be heard at any time of the year.
Habitat: This widely distributed frog is found in woodland, rainforest, farmland, heathland and grassy areas. It is often noticed after rain

Limnodynastes dumerilii

70mm

and is commonly associated with dams, ditches and other bodies of still water.

Similar species: The prominent tibial gland distinguishes it from all frogs except *Limnodynastes terraereginae*, which can be recognised by the scarlet in the groin, and *Limnodynastes interioris*, which is distinguished by its reduced webbing.

Limnodynastes fletcheri

Barking Marsh Frog
Limnodynastes fletcheri
fletcheri is after H. Fletcher, the Australian zoologist

Key No. 58B

Description: This frog of up to 50mm is light grey or brown above, with scattered, irregular-shaped dark blotches and spots. There is usually

50mm

a pink or purplish patch on the hind side of the upper eyelid. The skin is smooth, or with low, rounded warts. The belly is smooth and white. The toes have basal webbing and slight fringes. It has prominent vomerine teeth behind the choanae.

Calls: A dog-like "rok" or "whuck". Males call from the water among vegetation after rain, particularly between October and March.

Habitat: The species is found in water-covered grassy areas and around the shores and banks of large western lakes and rivers. It may be found sheltering under rocks and logs, and in cracked mud and old yabby burrows.

Similar species: This species is often confused with *Limnodynastes tasmaniensis*, from which it can be distinguished by the pink/purple eyelids and irregular-shaped blotches on the back.

Giant Pobblebonk, Giant Banjo Frog
Limnodynastes interioris

interioris means "inner", as in Central Australia

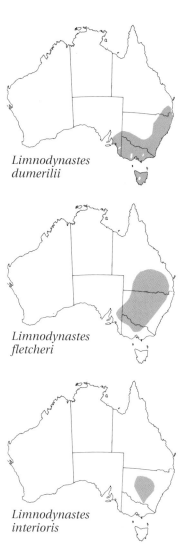

Limnodynastes dumerilii

Limnodynastes fletcheri

Limnodynastes interioris

Limnodynastes interioris
90mm

Limnodynastes ornatus

Limnodynastes ornatus
45mm

Key No. 59B
Description: This frog has a large snout, up to 90mm long, ranging from pale fawn to yellowish, to red-brown above. It usually has a few scattered blackish flecks and spots. There is usually a broad coppery-orange band down the side, with an irregular black band below it. There is a raised yellow or orange stripe from below the eye to below the tympanum. The belly is yellow, often with black flecks. The prominent tibial gland is usually copper or orange. The skin is smooth to finely granular above, and smooth on the belly. The toes are from a third to half webbed. There are prominent vomerine teeth behind the choanae.
Calls: A deep "plunk" or "thud", similar to *Limnodynastes dumerilii*. Males call while concealed in burrows in the banks of dams, pools and similar water bodies. They may also call while concealed in vegetation in the water. Males have been heard calling from August through to December, and again in March —

the summer break is probably due to lack of rain.
Habitat: It lives in woodland and open and disturbed areas. Dams and pools with marginal vegetation are a preferred breeding site.
Similar species: It can be distin-guished by its tibial gland from all species except *Limnodynastes dumerilii*, from which it is distin-guished by the greater expanse of webbing and the bright belly colouration, and *Limnodynastes terraereginae*, from which it is distinguished by the lack of scarlet patches in the groin and hind side of thigh.

Ornate Burrowing Frog
Limnodynastes ornatus
ornatus means "ornate"

Key No. 55A
Description: A variable, up to 45mm long frog, ranging from an almost uniform dark grey or brown to pale grey or brown with irregular

39

darker blotches. The limbs are barred or spotted with the same dark colouration. There is frequently a U- or butterfly-shaped pale patch on the back, behind the eyes. The belly is white and smooth, and the back has a scattering of small warts. The toes have a slight web. There are prominent vomerine teeth behind the choanae.

Calls: Rapidly repeated gulp sounding "unk.......:.unk.........unk". The males call while floating freely in the water after heavy rain.

Habitat: It ranges from wet sclerophyll forest in coastal areas through to dry, arid woodland.

Similar species: It can be distinguished from the other rotund NSW *Limnodynastes* by the absence of a tibial gland and from *Neobatrachus* by the pale metatarsal tubercle, and *Notaden* by its colouration.

Striped Marsh Frog

Limnodynastes peronii

peronii is after F. Péron, the French zoologist

Key No. 58A

Description: This up to 65mm long frog is one of the most common on the east coast. It is light brown or grey-brown above, with darker brown stripes. Some individuals have a pale stripe running down the back. The belly is white. The iris is golden above and dark brown below. There is a pale raised stripe running from below the eye to the base of the arm. The skin is smooth all over. The toes are almost free of web. There are prominent vomerine teeth behind the choanae.

Calls: A loud "tok" or "whuck" noise, similar to the sound of a tennis ball being struck. The males call while floating in water, or when on land near the water's edge. The males can be heard throughout the year. It is an adaptable species and often calls from suburban fish ponds, or even from swimming pools.

Habitat: It can be found along the coast and in the ranges of eastern Australia. It appears tolerant of polluted water.

Similar species: It can be distinguished from *Limnodynastes*

*Limnodynastes
peronii*

*Limnodynastes
peronii*
65mm

Limnodynastes salmini   75mm

tasmaniensis and *Limnodynastes fletcheri* because it has stripes instead of spots. It can be distinguished from *Limnodynastes salmini* by its lack of pink/orange stripes and the two-coloured iris.

Pink Striped Frog, Salmon Striped Frog
Limnodynastes salmini

salmini is after Salmin, whose identity is unknown.

Limnodynastes salmini

Key No. 57A

Description: A rather large frog at up to 70mm, it is similar to the previous species (*Limnodynastes peronii*). It is brown or grey-brown above, with scattered dark brown spots and blotches. There are usually three pink, orange-red or red-brown stripes — two on each side, running from shoulder to groin, and the third running down the back. The belly is white, while the sides, groin and edges of the thigh are mottled black and white. The iris is golden above and below the pupil. The skin is smooth all over. The toes have a rudimentary web. There are prominent vomerine teeth behind the choanae.

Call: An "unk-unk-unk" call. The males call from matted vegetation around the edges of inundated ponds, ditches and marshes. The males have been heard calling from September to April, after heavy rain.

Habitat: This species apparently spends much of the year buried underground. After heavy rains, it is usually found under logs, rocks and loose bark, close to its breeding ponds.

Similar species: The pink/orange stripes are sufficient to distinguish this species from *Limnodynastes peronii*, *Limnodynastes tasmaniensis* and *Limnodynastes fletcheri*.

Spotted Marsh Frog
Limnodynastes tasmaniensis

tasmaniensis means "from Tasmania".

white. The iris is golden above and below the pupil. The toes are fringed and slightly webbed. There are prominent vomerine teeth behind the choanae.

Call: A machine gun-like "uk-uk-uk", repeated at intervals. In the southern part of Victoria, this call is reduced to a single "pok". The males call from the water in either concealed or exposed sites, throughout the year and particularly after rain.

Habitat: Like *Limnodynastes peronii* this is a very adaptable species and is

Limnodynastes tasmaniensis

Limnodynastes tasmaniensis

45mm

Key No. 58C

Description: A smallish frog, reaching 45mm. It is light brown to rich olive green above, with a series of irregular darker spots and blotches. Some specimens have a white, yellow or pinkish stripe running down the back. There is a raised pale stripe running from below the eye almost to the base of the arm. The legs are blotched like the body. The skin is smooth or with low warts above, and it is smooth on the belly. The belly is

often one of the first frogs to take advantage of new dams, ditches and water-covered areas on disturbed ground. It can be found in woodland, shrubland and grassland from the coast to the interior. It is usually found under cover near water by day.

Similar species: This species can be distinguished from *Limnodynastes fletcheri* by the more regular shape of its blotches and by the lack of pink/purple eyelids. It can be distin-

Limnodynastes terraereginae

Limnodynastes
terraereginae

Limnodynastes terraereginae   75mm

guished from *Limnodynastes peronii* by the lack of divided eye colour and the spotted (not striped) pattern. It can be distinguished from *Limnodynastes salmini* by the lack of three pink stripes.

Northern Banjo Frog, Northern Pobblebonk
Limnodynastes terraereginae

terraereginae means "from Queensland"

Key No. 59C

Description: This is a large, up to 75mm long, stout burrowing frog. It is grey or brown above, with occasional darker flecks and blotches. A few specimens may have a dorsal stripe. There is a cream or reddish-orange raised stripe from below the eye to the base of the arm. Above this, a dark stripe runs from the snout through the eye, ending at the shoulder. There is often a reddish-orange stripe on each flank and a reddish patch on the upper arm. The belly is white or pale yellow. The groin and hind side of the thigh have red or scarlet markings. The skin is smooth or slightly granular above, and smooth below. The toes have basal webbing. There are prominent vomerine teeth behind the choanae.

Call: A rather high-pitched, short "dunk" or "bonk". The males call while concealed in water — usually from holes in the banks of dams, flooded ditches and similar places — from October to May, after rains.

Habitat: They are found in forest, woodland and cleared land. They are usually found near permanent water after rains.

Similar species: This species is readily distinguished from all similar species in south-east Australia by its scarlet groin and thigh markings.

GENUS

Mixophyes

*There are five species known in this genus at pre-
sent, all of them restricted to eastern Australia.
Four of the species are found in the area of this
guide. All these frogs have banded legs, obviously
webbed feet and vertical pupils. The species are
usually found in wet, forested situations like
Antarctic Beech forest, wet sclerophyll forest and
rainforest. The males call on land near water, and
spawning occurs there.*
Mixophyes *means "slimy kind".*

Great Barred Frog
Mixophyes balbus (27A)
Mixophyes fasciolatus (28B)
Mixophyes fleayi (28C)
Mixophyes iteratus (28A)

balbus means "stuttering";
fasciolatus means "striped";
fleayi is after the Australian
zoologist David Fleay;
iteratus means "repeating"

Key Nos. 27B, 28A-C
Description: These are all very similar species with similar habitats, and
so they are grouped together. They
are all large, ranging from 80mm to
115mm as adults (*M. iteratus* being
the largest). The differences between
the species are listed under "Calls"
and "Similar species". They range
from yellowish-grey to almost black
above, with darker spots and mottling. The legs in all species are

*Mixophyes
fasciolatus*

*Mixophyes
fleayi*

barred or banded. There is a thin
dark stripe running from the snout
through the eye, and above the tym-

Mixophyes
balbus

Mixophyes
balbus
80mm

Mixophyes fleayi   80mm

Mixophyes
fasciolatus
80mm

panum. The skin varies from smooth to finely granular above, and it is smooth below. The bellies of all species are white or pale yellow. The toes are three-quarters to fully webbed. There are prominent vomerine teeth in front of the choanae and there are maxillary teeth as well.

Calls: *M. balbus* has a short "op .. op.op", and an "a.a.a.ah", the latter being heard when more than one male is calling. *M. fasciolatus* has a deep, harsh "wark".*M. fleayi* has both an "op..op..op.." (solitary), and an "a.a.a.a.a.ah" (chorus). *M. iteratus* has a deep, guttural grunt. All species call from leaf litter along the banks of creeks and streams. The eggs are also laid there, to be washed into the water later by heavy rains. All known breeding seems to be in late spring and early summer.

Habitat: All species seem restricted to wet sclerophyll forest and rainforest, including Antarctic Beech forest. They are usually found fairly close to permanent running water.

Similar species: The genus *Mixophyes* can be readily distin-

*Mixophyes
iteratus*

Mixophyes iteratus   115mm

guished from *Lechriodus fletcheri* by the lack of skin ridges and more extensive webbing.

M. balbus has narrow and indistinct dark crossbars on the hind limbs. There are no conspicuous black spots or blotches on the sides. The iris is pale above and dark below and there is usually a pale blue crescent in the top half of the eye. The toes are three-quarters webbed. The belly ranges from white to pale yellow.

M. fasciolatus has well defined dark crossbars on the limbs, which broaden towards the back of the thigh to form triangular patterns. There is a series of conspicuous black spots and blotches on the side and the upper lip is pale creamy-white, without dark blotches. The iris is also a uniform dark brown. The toes are three-quarters webbed.

M. fleayi, has well defined dark crossbars on the limbs, which broaden towards the back of the thigh to form triangular patterns. There is a series of conspicuous black spots and blotches on the side and the upper lip is brownish, often with one or more dark purplish-brown blotches. The iris is either pale blue, or silvery in the top half. The toes are three-quarters webbed.

M iteratus has dark crossbars on the limbs of equal width to the pale spaces between the bars. There is a broad area of scattered dark spots on the sides. The iris is pale golden in the upper half, and darker in the lower half, and the toes are fully webbed.

GENUS

Neobatrachus

There are at least nine species in this widely distributed genus. They are all plump burrowing frogs with vertical pupils, and they are so similar that they can often only be identified by their calls and distribution, or by genetic analysis.
The three species in the guide can all be separated by external features, such as the amount of pigmentation in the metatarsal tubercle and the presence or absence of loose groin skin.
Neobatrachus means "new frog".

Trilling Frog
Neobatrachus centralis
centralis means "central"

Key No. 60C
Description: This species grows to 55mm. It is pale brown or yellow above, with small, irregular darker

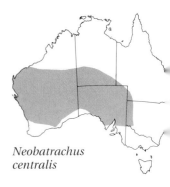

Neobatrachus centralis

Neobatrachus centralis
55mm

patches and usually a pale stripe down its back. It is smooth above, except in the case of breeding males, which become spiny. The belly is smooth and white. The toes are fully webbed with deep indentations between the toes. The metatarsal tubercle is edged with black or brown. The vomerine teeth are broad and situated between the choanae.

Call: A long, high-pitched trill. The males call while floating in water, usually after heavy summer rain when they congregate at water-filled ditches and claypans.

Habitat: They are found on clay and loam soils, apparently replacing *Neobatrachus sudelli* in the west of the state.

Similar species: It can be distinguished from other burrowing frog genera (*Limnodynastes*, *Heleioporus* and *Notaden*) by the combination of a pigmented metatarsal tubercle, a vertical pupil and fully webbed toes. It can be distinguished from *Neobatrachus pictus* and *Neobatrachus sudelli* by the

metatarsal tubercles, which are only pigmented on the edge, rather than completely black as they are in the other two species.

Mallee Spadefoot, Painted Burrowing Frog
Neobatrachus pictus

pictus means "painted"

Key No. 60B
Description: This is a plump, burrowing frog, reaching 50 mm long. It ranges from grey through to yellow above, with dark brown or olive green blotches. Often there is a thin yellow or cream stripe running down the back. The belly is white. The skin varies from warty to spiny above, with many of the warts having yellow tips. The belly is smooth. The toes are webbed, and the metatarsal tubercle is black. The frog has broad vomerine teeth behind the choanae.
Call: It gives a long trill. The males usually call while floating in the water, generally after heavy summer rains.

P.ROACH

Neobatrachus pictus

Neobatrachus pictus

50mm

Habitat: It is found in woodland, mallee, grassland, farmland and cleared areas. It is usually found after rain near flooded dams, ditches and claypans.

Similar species: It is most similar to *Neobatrachus sudelli*, which has loose, baggy groin skin (lacking in *Neobatrachus pictus*). The wholly black-pigmented metatarsal tubercle is not found in other burrowing frogs. *Neobatrachus centralis* has a pigmented metatarsal tubercle, but this is only dark-edged, not completely black.

Painted Burrowing Frog
Neobatrachus sudelli

sudelli is after Sudell, whose identify is unknown.

Key No. 60A

Description: A highly variable, up to 40mm long, species. Its colour may be grey, yellow or red-brown

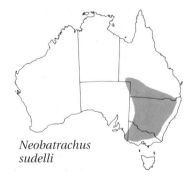

Neobatrachus sudelli

Neobatrachus sudelli

40mm

above, with irregular darker spots and blotches. There may be a pale stripe down the back, and the belly is white. The skin is rough and warty on the back, but smooth on the belly. The skin in the groin is loose and baggy, extending from the groin to the knee. The toes are fully webbed, but there is a deep indentation in the webbing between the toes. The metatarsal tubercles are completely black. The vomerine teeth are broad and situated between the choanae. **Call**: A short trill. The males call while floating in water, usually after heavy summer rains when ditches, claypans, ponds and dams fill with water.

Habitat: It lives in woodland, shrubland, grassland and disturbed areas. **Similar species**: It can be distinguished from burrowing frogs of different genera (e.g. *Limnodynastes, Heleioporus* and *Notaden*) by the combination of a black metatarsal tubercle, a vertical pupil and fully webbed toes. It can be distinguished from *Neobatrachus centralis* by its completely black (as opposed to dark edged) metatarsal tubercle and from *Neobatrachus pictus* by the baggy groin skin.

GENUS

Notaden

*There are four warty, plump, burrowing frogs in this genus, all with horizontal pupils, but only one is found in the area covered by this guide. This also happens to be the most colourful of the species. They all seem to feed on ants and termites, but they will take other small invertebrates as well. The warty backs have glands which produce a sticky yellowish poison.*Notaden *means "back gland".*

Crucifix Frog, Holy Cross Toad
Notaden bennettii
bennettii refers to the museum collector, G. Bennett.

Key No. 55B
Description: A very distinctive frog, reaching 50mm and unlikely to be confused with other species, as it is olive, yellow, or green above. A cross-shaped pattern of black, red, white and yellow spots and warts is on the back, hence its common name. There are usually scattered yellow, black and white spots on the sides. The skin is warty above, and smooth on the belly. The toes are slightly webbed. The frog has a large, pale, shovel-shaped inner metatarsal tubercle, and prominent vomerine teeth.

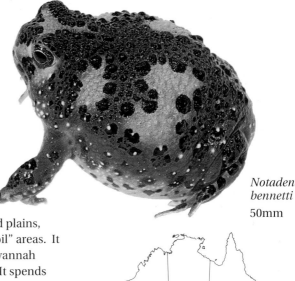

Call: An owl-like "whooo", rising slightly in inflection. The males call while floating freely in the water, where they are usually seen and heard only after heavy summer rain.

Notaden bennetti
50mm

Habitat: It is found on the slopes and plains, particularly in "black soil" areas. It can also be found in savannah woodland and mallee. It spends most of the year buried underground and feeds mainly on small black ants.

Similar species: Its striking colour and pattern, plus the pale metatarsal tubercle, distinguish it from all other frogs in the area.

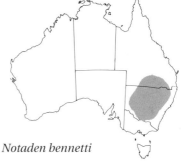

Notaden bennetti

GENUS

Paracrinia

There is one species in the genus, restricted to eastern New South Wales and eastern Victoria. Its toes are fringed, but unwebbed, it has a horizontal pupil, and a brown belly with pale spots. There are orange-red patches in the armpit and groin, and also on the backs of the thighs.
Paracrinia *means "similar to* Crinia".

Haswell's Froglet
Paracrinia haswelli

haswelli is a reference to the zoologist, W. Haswell

Key No. 16A
Description: An up to 32mm long frog, it is light grey-brown to brown above, with irregular darker flecks. A pale stripe may run down the back. There is a black band from the nostril through the eye and the tympanum, down to the sides, where it may break up into a series of blotches. The belly is light brown with paler

Paracrinia haswelli
32mm

Paracrinia haswelli

flecking. There are bright red patches at the base of each arm, in the groin and on the backs of the thighs. The skin is smooth, or with a few low tubercles above, and slightly granular on the belly. The toes are unwebbed, but fringed. The frog has both vomerine teeth and maxillary teeth.

Call: A short "ank", slowly repeated. The males call either while floating in concealed vegetation, or while concealed in litter and vegetation on the bank. They have been recorded calling from August to March.

Habitat: It lives in wet and dry scle-rophyll forest and heathland. It is usually associated with water such as creeks, ponds, dams, ditches and swamps.

Similar species: Although superficially resembling *Crinia*, *Uperoleia* or *Pseudophryne*, it can be readily distinguished from these by its combination of smooth belly, belly pattern and red groin patch.

Philoria

*This genus contains four species at present,
and all are found in the area covered by this guide.
There is some controversy about this genus, with
some people arguing that only* P. frosti *should be in
the genus* Philoria, *and that the others should be in
the new genus* Kyarranus. *They are listed in this
way in some books, but this guide follows
Cogger (1992).
All the species lay their eggs in damp burrows.
Unusually, the tadpole stage passes within the eggs
in the burrow, so that the young "hatch" as fully
developed small frogs, or advanced tadpoles which
do not feed, but rapidly develop limbs. These frogs
all have unwebbed toes and a horizontal pupil.*
Philoria *means "mountain lover".*

Baw Baw Frog
Philoria frosti

frosti is after C. Frost, the naturalist

Key No. 2B
Description: This frog reaches
45mm long, making it the largest of
its genus. It is dark brown on the
upper surface, with a scattering of
darker flecks and lighter blotches.
The frog's prominent parotoid gland
is dark brown or black. There are
often yellow or cream markings on
the head between the eyes.
Occasionally, this yellow colouration
continues as a broad band down the
back. The belly is cream or yellowish, with much brown flecking. The
skin is rough, with low warts and
tubercles on the upper surface, and
the belly is smooth. The toes are free
of webbing. There are vomerine
teeth behind the choanae, and maxillary teeth as well.
Call: The species has been recorded
making a variety of calls, usually
either "clunk", or "uk . . . uk . . . uk",
repeating at intervals. The males call
in late spring from burrows in damp
ground and sphagnum moss.
Habitat: So far, this frog is only
known from the higher slopes of Mt
Baw Baw in Victoria. It has been
found under stream-side rocks and
logs, and also in sphagnum bogs in
heath and grass areas.
Similar species: Due to its colouration and distribution, it is unlikely to
be confused with any other frog.

Philoria frosti

Philoria frosti
45mm

Yellow-bellied Mountain Frog
Philoria kundagungan

kundagungan is an Aboriginal word for mountain frog

Key No. 53C
Description: This frog is about 30mm long. On the upper surface, it is variable, yellow or red to black, occasionally with two black V-shaped blotches above each groin. The belly is a bright yellow colour. The frog's skin is smooth, both above and on the belly. Its toes are not webbed. It has vomerine teeth behind the choanae, and maxillary teeth as well.
Call: A deep, slow "orp". The males call from burrows during late spring and through the summer.
Habitat: The frog inhabits mountainous forest country, where it is found in small creek beds associated with water-covered leaf litter and mud.
Similar species: The bright yellow belly distinguishes this species from all other *Philoria* and similar species.

Loveridge's Frog
Philoria loveridgei

loveridgei is after A. Loveridge, the American zoologist

Key No. 53B
Description: This frog reaches a length of 30mm . It is almost uni-

R.W.G.JENKINS

Philoria kundagungan

Philoria kundagungan
30mm

Philoria loveridgei

Philoria loveridgei
30mm

form brown above, with a dark band from the snout through the eye to the shoulder. The belly is white with a few brown flecks, and the throat is heavily dotted with dark brown. The skin is smooth or with a few low warts and ridges above, and smooth on the belly. There is a fold of skin above the tympanum. The toes are free of webbing. There are vomerine teeth in oblique rows behind the choanae, and the frog also has maxillary teeth.

Call: A guttural "orp" or "ork" sound. The males call from underground burrows. They are thought to call from late spring to early summer.

Habitat: It inhabits rainforests including Antarctic Beech forest and wet sclerophyll forests, above 750m altitude. It is usually found in burrows in moist soil or moss.

Similar species: This frog is unlikely to be confused with others due to its habitat requirements and colouration. *Philoria sphagnicolus* has a distinctive dorsal pattern not found in this species, and it is larger. *Philoria kundagungan* has a bright yellow belly, and *Taudactylus diurnus* has fringed toes with obvious discs.

Sphagnum Frog
Philoria sphagnicolus

sphagnicolus means "moss dwelling"

Key No. 53A.
Description: This species reaches 35mm. It ranges in colour from cream through yellow and red to black above, usually with darker flecks and patches associated with low tubercles. A dark band runs from the snout through the eye to the shoulder. Another is along the side, while a third band runs from the groin to the back sometimes joining with the band on the other side in the middle of the back to form an inverted "V". Some individuals have a cream-coloured broad stripe down the back. The limbs have faint dark bars. The belly is white, while the throat is flecked with brown. The skin is smooth, or with a few low warts and folds above, and smooth on the belly. The toes are unwebbed. There are vomerine teeth present as two oblique rows behind the choanae, and there are maxillary teeth.

Call: A low growl, "creeerk" or "gur-rrr". The males appear to call from

Philoria sphagnicolus   35mm

*Philoria
sphagnicolus*

burrows in similar situations to the previous species, *Philoria loveridgei*.
Habitat: It lives in rainforests, including Antarctic Beech forest, wet sclerophyll forests, and sphagnum moss beds. It is usually found in burrows either in moss or damp soil, or under rocks and logs.
Similar species: This frog is unlikely to be confused with other species due to its habitat requirements and distinctive dorsal pattern.

GENUS

Pseudophryne

Members of this genus can be found in all states, but not in the Northern Territory. They are mainly restricted to damp areas.
They all have strongly marbled bellies, usually patterned in black and white, and the calls of the species are similar. The toes are unwebbed in these frogs, and they crawl rather than hop. Most of the species which have been studied lay their eggs in damp burrows, and the young hatch out in an advanced stage when the burrows flood with water.
Pseudophryne *means "false toad".*

Red-crowned Toadlet
Pseudophryne australis

australis means "southern"

Key No. 23A
Description: This colourful frog is up to 30mm long and dark brown to black above, often with a red "wash", or scattered red flecks. There is an orange or red triangle on the head and a stripe of the same colour on the lower back. The base of each

and may be found beside temporary creeks, gutters and soaks within this area, and under rocks and logs. It breeds in damp leaf litter where the male remains in the vicinity; the eggs hatch when the tadpoles are already advanced in their development and the breeding site is inundated with heavy rain.
Similar species: Due to its colour pattern and distribution, this species cannot be confused with any other species.

Pseudophryne australis

Pseudophryne australis 30mm

arm is white. The belly is strikingly marbled with black and white. The skin is smooth to warty above, and smooth on the belly. The toes are unwebbed. There are no vomerine or maxillary teeth.
Call: All species of *Pseudophryne* have similar calls — a grating "ark" or "squelch". The males call from concealed sites on land and in burrows throughout the year. As the species is frequently found in colonies, several will usually be heard answering each other.
Habitat: The species seems to be restricted to Hawkesbury Sandstone

Bibron's Toadlet
Pseudophryne bibronii

bibronii is after G. Bibron, the French zoologist

Key No. 23B
Description: This frog reaches 30mm in length. It is brown to black above, with darker flecks and occasionally reddish spots. There is often a bright yellow spot on the vent, and sometimes a yellowish stripe on the lower back. The base of the arm is yellow. The belly shows a bold black and white mottling. The skin is

Pseudophryne bibroni 30mm

smoothly warty above, and smooth or slightly granular on the belly. The toes are unwebbed. There are no vomerine or maxillary teeth.

Call: As with other *Pseudophryne* species, the call is a grating "ark" or "squelch". The males call throughout the year, even in winter temperatures as low as 4°C. The males call from burrows in damp soil or while concealed in damp leaf litter, under rocks, or within grass clumps.

Habitat: It is found in forest, heathland or grassland, usually singly, unless it is in breeding aggregations. The eggs are laid in damp leaf mould or burrows under rocks and logs, and they hatch when rain floods the burrows.

Similar species: It may be readily distinguished from *Pseudophryne dendyi* only by limb length. The extended hind limb of *Pseudophryne*

Pseudophryne bibroni

bibronii, moved forward, will reach the tip of the snout or beyond, while it will not reach the snout in *Pseudophryne dendyi*. In other words, the total leg length from foot to hip is as long as the body, or longer in *Pseudophryne bibronii*, but shorter in *Pseudophryne dendyi*. It may be distinguished from all other species by its dorsal and ventral pattern. *Pseudophryne major* is also similar, but has yellowish marbling on its belly.

Pseudophryne coriacea

Pseudophryne coriacea
35mm

Red-backed Toadlet
Pseudophryne coriacea

coriacea means "leathery"

Key No. 23C
Description: This frog is up to 35mm long. It is a rich brown to bright red above, and may have some darker flecks. The sides of the body and head are black, while the base of the arm may be white. The belly is boldly marbled black and white. The skin is smooth, with occasional low, scattered warts, and it is smooth or faintly granular on the belly. The toes are unwebbed. There are no maxillary or vomerine teeth.
Call: As with other *Pseudophryne* species, the call is a grating "ark" or "squelch". The males call from shallow burrows and other cover.
Habitat: It lives in sclerophyll forest and flat marshy areas, where it is usually associated with leaf litter and logs. The eggs are laid in burrows in moist earth and they hatch when they are inundated by rain.
Similar species: Due to its colouration and the strong demarcation between the stripe down the side, the colouration of the back and the black and white belly, it is unlikely to be confused with any other species in south-east Australia. It can be distinguished from *Assa darlingtoni* by its belly colouration.

Corroboree Frog
Pseudophryne corroboree

corroboree — after "corroboree"

Key No. 23D
Description: This distinctive frog reaches 30mm long. It is bright yellow or greenish-yellow above, alternating with black stripes. The belly is black and white, or black and pale yellow. The skin is slightly granular, with longitudinal ridges above, and smooth on the belly. The toes are unwebbed. There are no maxillary teeth or vomerine teeth.
Call: As with other *Pseudophryne*, the call is a harsh "ark" or "squelch" sound. The males call from burrows, especially in sphagnum bogs, where they can be heard from November to

59

Pseudophryne corroboree
30mm

January. The breeding season is restricted by the alpine climate.
Habitat: It lives in grassy marshland, or under logs and vegetation beside creeks in sclerophyll forest. Sphagnum bogs are its preferred breeding habitat.
Similar species: None: the Corroboree Frog could not be confused with any other frog.

Dendy's Toadlet, Southern Toadlet
Pseudophryne dendyi

dendyi is after Dendy, the English zoologist
Key No. 20B
Description: This frog reaches

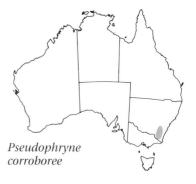

Pseudophryne corroboree

35mm in length. It is usually a dark brown or black with bright yellow patches on the upper arm, vent, back edge of the thigh and sometimes between the eyes, with a stripe running down the lower back. The belly is a bold black and white marbling. The skin is smoothly warty above, and smooth or slightly granular on the belly. The toes are unwebbed, and there are no vomerine or maxil-

*Pseudophryne
dendyi*
35mm

*Pseudophryne
dendyi*

-lary teeth.

Call: As with other *Pseudophryne*, the call is a grating "ark" or "squelch". The males call from low burrows and other cover. They have been heard calling from December to April.

Habitat: It is found in forest, heathland or grassland. It is usually found singly, unless in breeding aggregations. The eggs are laid in breeding burrows and hatch when rain floods the burrows.

Similar species: This species is very similar to *Pseudophryne bibronii*. It can only readily be distinguished by

the shorter hind legs. The extended leg, from foot to hip, is shorter than the body length in this species, but the same length as the body, or longer, in *Pseudophryne bibronii*. Its dorsal and ventral colouration distinguish it from all other frogs.

Great Toadlet, Major Toadlet
Pseudophryne major

major means "larger"

Key No. 22A
Description: To some extent, the common name is an exaggeration, as the adult only grows to 40mm long, but this still makes it Australia's largest toadlet. The upper surface is brown, with darker marbling, while the belly is black or brown, with yellowish marbling. The skin has scattered warts and ridges on the

back, while the belly is smooth. The
toes are unwebbed, and there are no
maxillary or vomerine teeth. This
species has a rather pointed snout.
Call: An "ark" or a "squelch" sound,
which has a grating quality. Little is
known about this species, but it is
thought to be similar to

*Pseudophryne
major*

Pseudophryne major   40mm

*Pseudophryne
semimarmorata*

Pseudophryne semimarmorata
30mm

Pseudophryne bibronii.
Habitat: The toadlet lives in damp or boggy areas in forest or heathland, and it probably burrows in damp leaf litter or under rocks and logs to breed.
Similar species: The yellowish marbling on the belly and its pointed snout distinguish the Great Toadlet from Bibron's Toadlet (*Pseudophryne bibronii*).

Southern Toadlet, Orange-throated Toadlet
Pseudophryne semimarmorata

semimarmorata means "half marbled"

Key No. 21C
Description: This toadlet is up to 30mm long. On the back, it is olive to dark brown, with darker flecks. The chest area of the belly has black and white marbling, while the throat, the underside of the legs and the lower half of the belly are bright orange, red, yellow, or flesh coloured. The skin has numerous low warts above, and in the females it is smooth below, while the males have a granular belly. The toadlet has no vomerine or maxillary teeth.
Call: Similar to other *Pseudophryne*, the call is a grating "ark" or "squelch". The males call from shallow burrows near boggy ground or water, and they have been heard calling in late summer and autumn.
Habitat: It is found in sclerophyll forest, woodland, heaths and grasslands. It is usually found under litter, logs and rocks in damp areas.
Similar species: The colouration makes it unlikely that this frog could be confused with any other.

GENUS

Rheobatrachus

There are only two species known in this genus, both restricted to three mountain ranges in eastern Queensland. Neither had been seen for several years when this book went to print, and they may be extinct.
These frogs have a remarkable method of reproduction: the tadpoles develop in the stomach of the female, and are later regurgitated as metamorphs (young frogs). Both frogs are aquatic, with fully webbed toes and short noses. The eyes and the nostrils are both directed upwards.
Rheobatrachus *means "stream frog."*

Southern Gastric Brooding Frog
Rheobatrachus silus

silus means "pug-nosed"

Key No. 4B
Description: This aquatic frog can grow up to 50mm long. The back colour varies from olive brown to almost black, with occasional darker blotches. There is a dark streak from the base of the arm, and there may be dark bars on the limbs. The belly is white, although the underside of the limbs may be yellow. The fingers, toes, and webbing are mottled with light and dark brown. The skin is shagreened or finely granular above, and smooth on the belly. The toes are fully webbed, and there are no discs present, but the tips of the toes may be slightly swollen. The pupil of the eye is vertical. There are no vomerine teeth, but the frog has well-developed maxillary teeth.

Rheobatrachus silus   50mm

Rheobatrachus silus

Call: An "eeeehm . . . eeeehm" sound, with an upward inflection. The males call from the water's edge during summer.
Habitat: It is found in rocky mountain creeks, where it hides beneath the rocks on the stream bed in the daytime. This frog has suffered a massive decline in recent years, and may be extinct.
Similar species: The habitat require ments, colour pattern and mottled full webbing distinguish this frog from all other species.

Taudactylus

This genus contains a number of small and agile frogs, which are restricted to eastern and north-eastern Queensland. Only one species occurs in the area covered by this book.
The toes may have basal webbing or fringing. Due to a T-shaped structure on the end of the toe (a terminal phalange), most species have noticeably expanded tips to their toes. From this, they superficially resemble members of the genus Litoria.
These frogs are mainly restricted to mountainous areas. They are often associated with running water and are active during the day, giving them their name of "Torrent Frog" or "Day Frog".
Taudactylus *means "T-fingered".*

Day Frog, Mount Glorious Torrent Frog
Taudactylus diurnus

diurnus means "daytime"

Key No. 6B
Description: The frog is up to 30mm long and active by day, hence its first common name. In colour, it ranges from grey to brown, with darker mottling. There is usually a pale stripe running from the back of the eye to the base of the arm, and a dark, H-shaped mark on the "shoulder" area of the back. The arms and legs are banded with dark grey or black, and the belly is a cream or bluish colour. The skin is smooth or granular, and/or with a few warts above, while it is smooth on the belly. There are wedge-shaped toe pads, and the toes are fringed but unwebbed. There are

Taudactylus diurnis

no vomerine teeth, but it has maxillary teeth.
Call: A quiet "cluck-cluck-cluck-cluck" noise. The males call from beside streams in the summer months.
Habitat: It is found on the banks of flowing rocky mountain streams. It readily dives into the fast-flowing water if it is frightened, and it can cling beneath rocks until danger has passed. Like *Rheobatrachus silus*,

Taudactylus diurnis

30mm

this frog has declined drastically in numbers, and may be extinct.

Similar species: Although it is superficially like some species of *Crinia* and *Litoria*, this frog can be distinguished by the discs on the toes, combined with the lack of any webbing.

Uperoleia

This genus is found in every state except Tasmania. It is generally similar in appearance to Pseudophryne, *but can be distinguished by the prominent glands on its back, and the red, yellow or orange patches in the groin and behind the knee. Several of the species are similar enough to cause problems in identifying them, and there have been two revisions of the genus since 1980 (see the References section on page 109). The pupil is horizontal, but rhomboidal (diamond-shaped), while the toes may have fringes and/or basal webbing.* Uperoleia *means "smooth back".*

Small-headed or Lumpy Toadlet

Uperoleia capitulata

capitulata means "small-headed"

Key No. 9A

Description: A smallish frog, reaching 28mm, with chocolate brown blotches on a grey and brown background. The enlarged parotoid, inguinal and coccygeal glands have cream markings. The belly is whitish, with slight brown stipples and patches. The groin and the backs of the thighs are scarlet. The skin is faintly or moderately rough and warty, and the belly is smooth. The toes are fringed but unwebbed. There are no vomerine or maxillary teeth.

Call: The call of this species is unrecorded.

Habitat: It has been found in mulga woodland and coolibah-lined water-holes. Common in the Bulloo River drainage system.

Similar species: This species is distinguished from other similar species

Uperoleia capitulata

Uperoleia capitulata   28mm

Uperoleia fusca   32mm

Uperoleia laevigata   29mm

by its enlarged parotoid, inguinal
and coccygeal glands with their
cream markings, plus the belly pattern. *Uperoleia rugosa* has yellow or
orange groin and knee patches, while
these are scarlet or red in *Uperoleia
capitulata*. *Uperoleia rugosa* has a
grey belly.

Uperoleia fusca

Uperoleia laevigata

Uperoleia martini

33mm

M. LITTLEJOHN

Dusky Toadlet
Uperoleia fusca

fusca means "dusky"

Key No. 10C

Description: A frog reaching 29mm, with skin varying from dark uniform slate to chocolate brown variegations on a grey or brown background. The parotoid glands are poorly to moderately developed, but the inguinal and coccygeal glands are not well developed. The belly may be cream with brown or slate speckling, and the belly colouring is uniform, not patchy. There is a red, orange, or yellow patch in the groin and on the backs of the thighs. The skin is faintly rough and warty above, and the

Uperoleia martini

belly is smooth. The toes are unwebbed, with slight fringes. There are no vomerine teeth, but the frog has maxillary teeth.

Call: A short, rasping "squelch" call. The males call from the base of grass clumps near water and have been heard in summer.

Habitat: It lives in open eucalypt

69

forest, tussock grassland and shrub-land.

Similar species: This species is distinguished from *Uperoleia laevigata*, *Uperoleia rugosa* and *Uperoleia capitulata* by its complete belly pigmentation, and from *Uperoleia tyleri* and *Uperoleia martini* by its poorly developed parotoid glands.

Smooth Toadlet
Uperoleia laevigata

laevigata means "smooth"

Key No. 10A
Description: A small frog up to 32mm, it is brown or olive brown above, with darker spots and blotches. There is usually a prominent pale triangular patch on the top of the head in front of the eyes, and there is a bright reddish or orange patch in the groin and behind each knee. The belly is pale, with a small amount of stippling, but it is never fully stippled or pigmented, giving a patchy appearance. The skin is granular or rough and warty above, and smooth below. The toes are unwebbed, but fringed. There are no vomerine teeth, but maxillary teeth are present.
Call: A "squelch" call. The males call on the ground, or in low vegetation near temporarily flooded grasslands, and have been heard calling from September to November.
Habitat: Dry forest and woodland, in grassy areas which can be covered by water after rain.
Similar species: This frog can be distinguished from the other species

of *Uperoleia* in the area by the prominent pale triangular patch on the head and the patchy belly pigmentation.

Martin's Toadlet
Uperoleia martini

martini is after A. Martin, the Australian zoologist

Key No. 10D
Description: This is a robust, up to 33mm-long frog. Its back is grey, with yellow and dark brown mottling, while its belly is a uniform brown, with white flecking. The groin and behind the knee are yellow. The skin has numerous tubercles on the back, but it is smooth on the belly. The parotoid glands are very well developed and conspicuous. The toes have no fringes or webbing. There are no vomerine teeth, but the frog has maxillary teeth.
Call: A prolonged and creaking "squelch" call. The males call from the edge of the water, either on the ground or in low vegetation and grasses inundated by water, usually after summer rains.
Habitat: It is found in coastal heath, dry sclerophyll forest, cleared land and grasslands. It is usually found under litter and logs and around inundated grassy areas.
Similar species: This species is most similar to *Uperoleia tyleri*, from which it can (with difficulty) be distinguished by its brown-with-white flecked belly (grey to blue-black in *Uperoleia tyleri*), and by the yellow mottling on the back (usually not

Uperoleia rugosa
32mm

found on *Uperoleia tyleri*), and by its call. This species can easily be distinguished from other similar species by the enlarged parotoid glands, but comparatively unenlarged inguinal glands and uniform belly pigmentation.

Uperoleia rugosa

Wrinkled, or Eastern Burrowing Toadlet
Uperoleia rugosa
rugosa means "wrinkled"

Key No. 9B
Description: A stout toadlet, up to 32mm long. The back is brown, with darker blotches and variegations, and there is usually a dark triangular patch on the head. The belly is grey and there are orange patches in the groin and behind each knee. The skin on the back is covered in tubercles, and both the parotoid and inguinal glands are enlarged. The belly is smooth, or slightly granular. The toes have a narrow fringe and are, at most, basally webbed. There are no maxillary or vomerine teeth.
Call: A "click" call. Males call from the edges of flooded depressions.and have been heard in July.
Habitat: It is found in dry forest, woodland and associated grassland. The frog is usually only seen after

Uperoleia tyleri
30mm

J. WOMBEY

heavy spring and summer rains. The males call from the water's edge, or from within cavities in grass tussocks and litter by the water's edge.

Similar species: This species is most similar to *Uperoleia capitulata*, from which it can be distinguished by the orange groin and knee colouration (scarlet or red in *Uperoleia capitulata*), and its grey belly. The enlarged inguinal glands, the belly colouration and the slight basal web to the toes distinguish it from all other species.

Tyler's Toadlet
Uperoleia tyleri

tyleri is after M. Tyler, the Australian zoologist

Key No. 10B

Description: A comparatively large species, up to 34mm long. A dark black-brown colour, with some orange-yellow spotting. The belly is blue-black, with white spots. The groin patch and thigh patch are yellow. The skin is faintly rough above, and smooth on the belly. The toes are unwebbed, with a slight fringe. There are prominent parotoid glands. There are no vomerine teeth, but maxillary teeth are present.

Call: A single short, pulsed or "squelch" call. The males call from the ground, or in litter close to the water's edge. They have been heard in September, November and January.

Habitat: It lives in dry forest, shrubland and grassland usually associated with areas that are flooded by rainwater.

Similar species: This species is most similar to *Uperoleia martini*, from which it can be distinguished (with difficulty) by its grey to blue-black belly, its lack of yellow mottling on the back, and its call. It can be distinguished from other *Uperoleia* by belly pattern and its well developed parotoid glands.

Uperoleia tyleri

FAMILY : HYLIDAE

These frogs are generally known as tree frogs, because two of the three Australian genera have adhesive toe discs which allow them to climb. This also gives a notched profile to the end of the toe. Only two of the genera, *Cyclorana* and *Litoria*, are found in the area covered by this book.

The species are widespread and diverse, especially in coastal and tropical areas. Most of the species which have been studied lay small, non-frothy eggs in water, where the tadpoles, with prominent pointed tails, develop normally.

Some arid zone species have rapid development, and some species which breed in running water have tadpoles with enlarged mouth parts. They use these to "suck onto" the rocks on the creek bed, and so avoid being swept away.

The name Hylidae comes from *Hyla*, a genus of foreign tree frogs. *Hyla* possibly means "barking".

GENUS

Cyclorana

This genus of burrowing frogs includes several species which are well adapted for the drier parts of Australia. Their distributions extend well into the arid zone, where drought conditions are frequent and prolonged. The members of this genus are found in all states except Victoria and Tasmania. The Cyclorana *frogs have no adhesive toe discs, and lack the notched profile to the toe, but they are otherwise similar in their skeletons and the shape of the tadpoles to the pattern found in the other Hylidae. The innermost finger (which corresponds to our thumb) is opposed to the other three fingers. The toes are webbed and the eye pupil is horizontal.* Cyclorana *means "round frog".*

Short-footed or Blotched Waterholding Frog

Cyclorana brevipes
brevipes means "short-footed"

Key No. 30C

Description: A robust burrowing frog, reaching 45mm long. Dark brown above, with light silver-brown blotches. There is usually a silver-brown stripe down the back. The body is white on the belly, and the backs of the thighs are dark brown, with a few lighter flecks. The skin is smooth or slightly granular above, and finely granular on the belly. The toes are less than a quarter webbed. This frog has vomerine teeth between the choanae, and maxillary teeth.

Call: A long, moaning growl.

Habitat: It lives in dry savannah woodland, usually seen after sum-

Cyclorana brevipes   45mm

mer rain near claypans.

Similar species: This species can be distinguished from *Cyclorana verrucosa* by the body colour and reduced webbing. As well, the metatarsal tubercle in *Cyclorana brevipes* is shorter than its distance to the tip of the innermost toe, while the tubercle is as long as, or longer than that distance in *Cyclorana verrucosa*. It can be distinguished from other *Cyclorana* species by the body colour pattern.

Cyclorana brevipes

Wide-mouthed Frog, Giant Waterholding Frog
Cyclorana novaehollandiae

novaehollandiae means "from New Holland"

Key No. 30A
Description: A large frog, up to 100mm long. Pale grey, brown, or yellowish above, occasionally with darker brown blotches and variegations. White on the belly, except for the throat which may be flecked with grey or brown. Juveniles may be bright emerald green, or with green patches. A dark stripe runs from the snout, through the eye and tympanum, extending down the sides, with a fold of skin above it (a dorso-lateral fold). A dark bar may be present under the eye. The backs of the thighs are grey or blue tinged, usually without variegations. The skin is smooth to tubercular above, and finely granular on the belly. The toes are one-third webbed. There are vomerine teeth behind the choanae, and maxillary teeth.

Call: A deep "waah" or "honk" sound. The males call from the banks and edges of clay pans, ditches and other bodies of temporary water. Normally only seen and heard after heavy summer rain.

Habitat: It occurs in woodland and associated grassland and claypan areas.

Similar species: A very distinctive species. Young specimens may be confused with *Cyclorana platycephala* or *Cyclorana verrucosa*. They can be distinguished from *Cyclorana platycephala* by the lack of fully webbed toes, and from *Cyclorana verrucosa* by the mottled or spotted colour of the back of the thigh in *Cyclorana verrucosa*.

Waterholding Frog
Cyclorana platycephala

platycephala means "flat-headed"

Cyclorana novaehollandiae

Cyclorana novaehollandiae

100mm

Key No. 31A
Description: This frog reaches 60mm in length. It can change colour, and varies from dull olive grey, or grey with green patches, to green with grey mottling. Often, a pale green stripe runs down the back. There are usually darker variegations, especially on the upper lip. The belly is whitish and the backs of the thighs are grey with darker flecks. The skin is smooth above, or with low warts, and the belly is smooth or granular. The toes are fully webbed. There are vomerine teeth between the choanae, and there are maxillary teeth.
Call: A slow "maw", somewhat like a motorbike. The males call from the water. They are usually only seen or heard after heavy summer rains.
Habitat: It is usually found in claypans, ditches, dams and temporary pools. The frogs are frequently found underwater, where they feed on other aquatic life, including smaller frogs and tadpoles.
Similar species: This species can be distinguished from most others by the full fleshy web on the hind feet, and by its colouration and distribution.

Warty Waterholding Frog
Cyclorana verrucosa
verrucosa means "warty"

Key No. 30B
Description: This frog grows to 45mm. It is grey-brown, olive green or green above, with irregular darker patches. There is usually a pale stripe down the back. There is a stripe from the snout, through the eye and tympanum, which merges

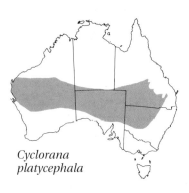

Cyclorana platycephala

Cyclorana platycephala

60mm

*Cyclorana
verrucosa*

Cyclorana verrucosa   45mm

frogs are usually only seen and
heard after heavy summer rain.
Habitat: It is usually found in and
around temporary ponds, ditches,
claypans and creeks in wooded or
open country. Found only after
heavy summer rains.

Similar species: This species can be
distinguished from *Cyclorana platycephala* by the lack of fully webbed
toes, and from *Cyclorana novaehollandiae*, which lacks any spotted or
marbled thigh colouration.
Cyclorana brevipes has toes which
are only quarter webbed and can be
distinguished by its different body
pattern and colouration, and by the
length of the metatarsal tubercle. In
Cyclorana verrucosa, the tubercle is
as long as, or longer than, the distance to the tip of the innermost toe.

into the body colouring at the
shoulder. The groin and hind side of
the thigh are grey-brown, with white
spots or mottling. The belly is
whitish, with brown mottling on the
throat. The skin is slightly to very
warty, and rough. The belly is granular. The toes are about one-third
webbed. There are vomerine teeth
between the choanae, and there are
maxillary teeth.
Call: A long, moaning growl. The

GENUS

Litoria

There are Litoria *species in every state, and also in Papua New Guinea. One species,* L. rubella, *is the most widely distributed species of frog in Australia. These frogs typically have adhesive discs on the fingers and toes. The cartilage between the disc and the rest of the toe gives a notched profile to the dorsal (upper) surface of the toe. Most species have obvious webbing between the toes, and all of them have horizontal pupils.* Litoria *means "shore" or "beach".*

Striped Burrowing Frog
Litoria alboguttata
alboguttata means "white speckled"

Key No. 31B
Description: This is a relatively slender, burrowing frog, up to 65mm long. It is brown, olive or green above, with darker blotches. There is

Litoria alboguttata

Litoria alboguttata   65mm

usually a pale yellow or yellow-green stripe down the back, and a dark streak runs from the snout, through the eye and the tympanum, breaking up down the flanks. This stripe has a skin fold above it. The backs of the thighs are dark, almost black, with large white or yellow spots. The belly is white, with some flecks of brown on the throat and chest. The skin of the back has scattered warts and ridges. The belly is granular, but the throat and chest are smooth. The toes are half webbed. Both vomerine and maxillary teeth are present.

Call: A rapid "quacking" call. The males call from around the grassy edges of temporary pools and ditches. They are often heard by day, and usually seen only after heavy summer rains.

Habitat: It lives in woodland and cleared areas. It is usually only seen around temporary pools and water-filled claypans. The species is active by day and night.

Similar species: Due to its distribution, it is unlikely to be confused with other species. It can easily be distinguished from the similarly coloured *Litoria aurea*, *Litoria raniformis* and *Litoria castanea* by the presence of a spade-like metatarsal tubercle for burrowing, and the lack of obvious adhesive toe discs.

Green & Golden Swamp Frog, Green & Golden Bell Frog
Litoria aurea

aurea means "golden"

Key No. 47A
Description: This frog can grow to 85mm. It ranges from dull olive to bright emerald green above, with varying amounts of brown or copper/bronze blotches; occasional specimens may be all green, or all bronze. There is a white or cream stripe running from above the nos-

Litoria aurea

Litoria booroolongensis   45mm

tril, over the eye and tympanum and continuing as a skin fold down the side. There is usually a darker stripe below this one and another pale stripe from below the eye to the base of the forearm. The groin and hind side of the thigh are a bright blue or blue-green. The belly is white. The skin is smooth or finely granular above, while the belly is coarsely granular. The toes are three-quarters to nearly fully webbed. There are vomerine teeth between the choanae, and maxillary teeth.

Call: A distinctive, four-part call, starting with a slow, drawn-out "craw-craw-crawk", followed by some short grunts, "crok-crok". It sounds rather like a distant motor-bike changing gears. The males call while floating, and they are usually heard from August through to January, although they have been heard at other times.

Habitat: It occurs in large perma-nent swamps and ponds with plenty of emergent vegetation, especially bulrushes. It is active by day and night. It will occasionally inhabit ornamental ponds and farm dams, where these occur close to the pre-ferred habitat. This species has declined in recent years.

Similar species: This species can be distinguished from similar species by its wart-free skin, conspicuous finger and toe discs, and lack of spotting or marbling on the hind side of the thigh. Its back colouration is also fairly distinctive.

Booroolong Frog
Litoria booroolongensis

booroolongensis means "from Booroolong"

Key No. 49B
Description: This frog grows to 45mm. It is dull grey or brown above, with paler spots and mottling, and the backs of the thighs are dark brown, with a few small pale spots. It is white or cream on the belly. The skin is smooth, or with scattered low tubercles. The chest and belly are granular, but the throat is smooth. Finger and toe discs are well devel-oped, but of moderate size. The toes are nearly fully webbed. The vomer-ine teeth extend from between to behind the choanae, and there are maxillary teeth.

Call: A slow, soft "quirk-quirk-quirk". The males call from rocks near or in the water, from August through summer.

Habitat: It is almost always associat-

Litoria aurea

Litoria booroolongensis

*Litoria
brevipalmata*
40mm

*Litoria
brevipalmata*

ed with rocky flowing streams in mountainous regions.

Similar species: This species vaguely resembles *Litoria lesueuri*, from which it is readily distinguished by the well-webbed feet and body pattern. *Litoria inermis* has a warty skin, and is not normally found in the mountains.

Green-thighed Frog
Litoria brevipalmata
brevipalmata means "short webbed"

Key No. 51C
Description: This attractive frog reaches 40mm long, and it is chocolate brown, with scattered black flecks. A dark stripe runs from the snout, through the eye and tympanum, and ends in the flank as a series of blotches. The upper lip may be white-edged, and this stripe continues to the base of the upper arm. The lower flanks are yellowish, with black flecking. The groin and backs of thighs are bright blue, green or blue-green, with black spots and blotches. The belly is white or pale yellow. The skin is smooth to slightly granular on the back, and granular on the belly. The toes are about one-third webbed, and the discs are moderate in size. There are vomerine teeth between the choanae, and maxillary teeth are also present.

Call: A continuous series of "quack" or "wok" sounds. The males call from around ponds and ditches of a semi-permanent nature. It has been recorded breeding in late spring and summer.

Habitat: It seems to be restricted to rainforest and wet sclerophyll forest.

Similar species: The thigh and groin colouration distinguish this species from any similar species.

Tasmanian Tree Frog
Litoria burrowsae
burrowsae is after M. Burrows, the collector of the first specimens

Key No. 34D, 35C
Description: This up to 55mm long frog seems only to be found in the Tasmanian highlands. It has either a light green or dark brown back: the

green form often has light brown patches, and the brown form has patches and flecks of green and light brown. There is a thin dark stripe, running through the eye, over the tympanum, widening as it continues towards the groin, and often becoming marbled with white along the flanks. The belly is a pinkish white, and the hind side of the thigh is pale brown, as is the groin. The skin of

the back is smooth, while the belly skin is granular. The toes are three-quarters webbed, the fingers are one-third webbed, and the frog has large toe discs. There are vomerine teeth between the choanae, and maxillary teeth.

Call: A goose-like "honk . . honk . . honk". The males call from grasses near water in the spring and late summer.

Habitat: These frogs have been collected from around the edges of alpine ponds, in grasses and reeds.

Similar species: The dis-

Litoria burrowsae

Litoria burrowsae

R.W.G.JENKINS

55mm

tribution and appearance make it unlikely this frog will be confused with any other. The only other Tasmanian tree frogs, *Litoria raniformis* and *Litoria ewingii*, do not have the pale brown thigh and groin colourations. As well, they have different back patterns and skin textures.

Litoria caerulea

Green Tree Frog
Litoria caerulea
caerulea means "blue"

Key No. 34A

Description: This frog reaches a length of about 100mm when fully grown. The upper surface is dark olive to bright green, but this can change over a period of an hour or so. At night, the frog can be a darker shade on top. The sides, and sometimes the back, may have a scattering of white spots or flecks. There may also be a series of white spots, or a stripe from the corner of the mouth to the base of the arm. The under surface is white. There is a skin fold running from the eye to the arm and the skin is smooth. The iris of the eye is golden, and the pupil is horizontal. The toe pads are large, the toes are three-quarters webbed, and the fingers are about one-third webbed. The vomerine teeth are between the choanae, and maxillary teeth are also present.

Call: A deep "crawk...crawk...crawk". It often calls from within retreats such as drainpipes, water-tanks and hollow branches. The males around breeding pools usually call from low branches etc., near the water's edge. A summer breeder, usually after rain.

Habitat: This species occurs in most habitats in Australia. It is, in fact, the second most widely distributed frog we have, being found in all states except Victoria and Tasmania. It is also found in southern New Guinea. A familiar frog to many people, it is often found around human habitation in places like shower blocks, toilets and water tanks. It will sometimes sit below outside lights at night to catch the insects which are attracted by the light.

Similar species: Due to its size, it is readily distinguished from most other green species in New South Wales. *Litoria chloris* is almost two thirds the size, and has a reddish iris and yellow upper arm, and no white spots or flecks.

New England Swamp Frog
Litoria castanea
castanea means "chestnut coloured"

Key No. 47C
Description: This frog can reach a length of 80mm. It is dull olive to bright emerald green on the back, with varying amounts of irregular bronze spotting and blotches. There is always a pale green stripe down the back. The tympanum is dark and conspicuous, and the dorso-lateral fold is pale cream and also conspicuous. There are scattered black spots on the back. The groin and hind side of the thighs are blue-green, with large yellow or cream spots, and the belly is white. The skin of the back is warty, and it is granular on the belly. The toes are fully webbed, with tiny toe discs — narrower than the toe

G.GRIGG

Litoria castanea

Litoria castanea 80mm

Red-eyed Tree Frog
Litoria chloris
chloris means "green"

Key No. 34C

Description: This frog reaches 65mm long. It is bright leaf-green to dark moss-green above. The backs of the thighs are purplish-red or brown, often with an iridescent tinge. The belly is either white or yellow, The upper arm, hands and feet are yellow, except for the outermost finger and toe, which are usually green, and there is also a small strip of green on the upper arm. The eye is golden and red in colour. The skin is finely granular or shagreened above, and granular on the belly. The tympanum is smooth. Hands and feet are almost fully webbed. There are vomerine teeth between the choanae, and maxillary teeth are present.

Call: Long moans, followed by soft trills. The males call from vegetation above breeding sites and in the shallows. They call after warm rains in spring and summer.

Habitat: It lives in wet sclerophyll

Litoria chloris

itself. The vomerine teeth are between the choanae, and there are maxillary teeth.

Call: Similar to, but not as complex as *Litoria aurea*. A series of droning grunts, rather like a distant motorbike. The males call while floating in the water. They can be heard most of the year.

Habitat: It is found in large permanent ponds, lakes and dams with an abundance of bulrushes and other emergent vegetation. This species has declined in recent years.

Similar species: It is very similar to *Litoria aurea* and *Litoria raniformis*, but it is readily distinguished from them because it has yellow spots in the groin and thigh, and tiny toe discs.

D.WHITFORD

Litoria cooloolensis   25mm

Litoria citropa   60mm

and rainforest, and it can also be found in associated flooded grasslands after summer rains. The species has been recorded breeding in neglected swimming pools.

Similar species: The eye colour distinguishes this frog from *Litoria caerulea* and *Litoria subglandulosa*. It is very similar to *Litoria gracilenta*, from which it can be distinguished by the green colouration extending onto the outside finger and toe plus thigh and upper arm. Also, *Litoria gracilenta* has the tympanum covered with granular skin, while in *Litoria chloris* this is smooth.

Blue Mountains Tree Frog
Litoria citropa
citropa means "lemon-coloured"

Key No. 41A

Description: This frog reaches 60mm in length. It is light to medium-brown above, usually with scattered darker flecks. A dark stripe runs from the snout through the eye above the tympanum to the groin. A thin pale stripe runs above this stripe. There is usually a bright green patch to be found on the side of the head, the upper and lower arm, the lower leg and flank. The armpit, groin, back of thigh and inner half of the foot are bright red-orange. The belly is white. The skin is smooth or finely granular above, with some scattered warts, and it is coarsely granular on the belly. The toes are half webbed, and the toe discs are large. There are vomerine teeth behind the choanae, and maxillary teeth.

Call: A harsh "warrk", followed by short trills "wrrk...wrrk...wrrk". The males call from vegetation and rocks close to permanent water in spring and early summer.

Habitat: It is usually found around rocky creeks and stream beds with associated woodland, or wet or dry sclerophyll forest.

Similar species: Due to its brown and green colour pattern, it cannot be confused with any other species,

Litoria citropa

Litoria dentata

Litoria dentata   45mm

except perhaps *Litoria subglandu-losa*, which has an indistinct tympanum.

Cooloola Tree Frog
Wallum Tree Frog

Litoria cooloolensis
cooloolensis mean "from Cooloolah"

Key No. 46C
Description: This is a small, slender frog, up to 25mm long. It is green above, usually with numerous brown spots and blotches. The tympanum is green, and the backs of the thighs are purplish-brown. The body is white or cream on the belly. The skin is smooth above, and granular on the belly. The toes are at least half webbed, but the fingers are webbed a third or less, and the toe discs are distinct. There are no vomerine teeth in this frog, but it has

maxillary teeth.
Call: The call is a "reek . . . pip", similar to the call of *Litoria fallax.* The frogs call from vegetation in or near water.
Habitat: It is often associated with water. The frog has been found in wallum country, lowland rainforest and around the freshwater lakes on Fraser Island.
Similar species: The species can be distinguished from *Litoria fallax* and *Litoria olongburensis* by its green tympanum and spotted body pattern.

Bleating Tree Frog,
Keferstein's Tree Frog

Litoria dentata
dentata means "toothed"

Key No. 36A
Description: This is a small but loud tree frog, up to 45mm long. It is creamish-brown to pale grey-brown above, with a broad dark brown band running from the head down the back. A dark stripe runs through the eye and tympanum, and down the side. The groin, the backs of thighs and the armpits are frequently yellow, particularly in males, and the

*Litoria
cooloolensis*

Litoria ewingii

belly is yellowish-white. The upper half of the iris is red. The skin is smooth above, with a few tubercles, and the belly is granular. The fingers are one-third webbed, while the toes are three-quarters webbed with large toe discs. There are vomerine teeth between the choanae, and maxillary teeth.

Call: A high-pitched bleat, almost painful in its pitch and volume. The males call from the ground close to water. They are usually only seen and heard after heavy spring and summer rain.

Habitat: It is often associated with coastal lagoons, ponds and swamps, particularly those with grassy edges. They can be found with difficulty beneath bark and stones near the breeding sites by day.

Similar species: The species is somewhat similar to *Litoria rubella*, from which it can usually be distinguished by the distribution and the broad dorsal strip. It can be distinguished from *Litoria verreauxii*, *Litoria ewingii*, *Litoria paraewingi*, *Litoria jervisiensis* and *Litoria revelata* by the presence of webbing between the fingers, and the red upper half of the eye. *Litoria rothii* does not have a dark band down its back.

Litoria fallax

Brown Tree Frog

Litoria ewingii

ewingii is after the naturalist, T. Ewing

Key No. 45B

Description: This frog is up to 45mm long, and pale to cream-brown on the back. It has a broad band running down the back with scattered darker flecks. A dark brown or black stripe runs from the snout through the eye and tympa-num to the shoulder, becoming indistinct on the flanks. There is a pale stripe below this dark stripe, running from the eye to the base of the upper arm. The belly is yellowish-white, cream or white. The backs of the thighs and the groin yellow to red-orange. There is no spotting or marbling in the groin or thighs. The skin is smooth or with low tubercles, and the belly is granular. The toes are half webbed, and there are small toe discs, only slightly wider than the toes. There are vomerine teeth between the choanae, and maxillary teeth.

Call: A whistling series of "weep..eep..eep..eep..eep" notes. The males call from the ground, low vegetation and while floating amongst vegetation. They have been heard all year round.

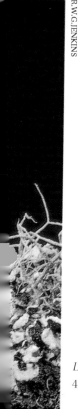

R.W.G.JENKINS

Litoria fallax   25mm

Litoria ewingii
45mm

Habitat: It can be found in most habitats, gathering to breed at farm dams, ponds, creeks and waterholes.
Similar species: It can be distinguished from *Litoria jervisiensis* and *Litoria dentata* by the small toe discs and the different groin and thigh colouration. It can be distinguished from *Litoria verreauxii* by the lack of black spots in the groin, and from *Litoria burrowsae* by its lack of green colouration. *Litoria paraewingi* can only be distinguished from it by the call and the distribution (compare the maps of distributions). *Litoria rubella* lacks any dorsal patch or band, and has a brown thigh colouration.

Green Reed Frog, Dwarf Tree Frog
Litoria fallax
fallax means "false"

Key No. 46A
Description: This slender frog is up to 25mm long and can be all green, green with fawn legs, or all fawn. A white stripe runs from below the eye to the base of the forearm. The backs of the thighs and the groin are orange to yellowish. The skin is smooth above, and the belly is granular. The toes are half to three-quarters webbed, and the toe discs are distinct. There are no vomerine teeth, but it has maxillary teeth.
Call: A ratchet-like "Reek...pip. Reek...pip...pip". The males call from emergent vegetation around swamps and dams. They have been heard from October to April.
Habitat: It is usually found not far from water and is common in emergent vegetation, especially bulrushes (*Typha*) spp., or swamps, lagoons and dams. It will also occupy nearby leaf axils, banana plants and bromeliads.
Similar species: Its elongate shape and white upper lip stripe should distinguish it from most other frogs that are similar in colour. Its lack of spots distinguish it from *Litoria cooloolensis*, which does not have an orange to yellowish thigh. The absence of brown throat speckling distinguishes *Litoria fallax* from *Litoria olongburensis*.

Freycinet's Frog
Litoria freycineti
freycineti is after L. Freycinet, the French navigator

Key No. 49A
Description: This frog reaches 45mm, and it ranges from pale grey-brown to dark brown above, with irregular darker blotches, warts and skin folds, usually in longitudinal rows. The "triangle" formed by the top of the head between the snout and eyes is pale with a darker central patch. A black stripe runs from the snout, through the eye and tympanum, to continue down the flanks.

Litoria freycineti

Litoria freycineti   45mm

There are at least two breaks in this stripe, one in front of the eye and one at the arm. The tympanum is almost completely ringed in white. A pale stripe runs below this dark stripe from the eye to the base of the upper arm. The belly is white or cream. The hind edge of the thigh is brown with large cream spots. The skin is smooth with low warts, tubercles and skin ridges, and the belly is granular, while the throat is smooth. The toes are almost fully webbed, with small toe discs a little wider than the toes. The vomerine teeth are between the choanae, and maxillary teeth are present.

Call: A rapid yapping or quacking call. The males call from the edges of ponds or from the ground nearby, in spring and summer after rain.

Habitat: This species can be found in a variety of habitats, particularly around temporary swamps.

Similar species: This species is similar to *Litoria nasuta*, from which it can be distinguished by the thigh pattern of brown and cream spots. It can be distinguished from *Litoria latopalmata* and *Litoria lesueuri*, which have no skin folds and ridges. It can be distinguished from *Litoria inermis* by its larger size and distinct body pattern.

Dainty Tree Frog, Banana Frog
Litoria gracilenta
gracilenta means "delicate"

Key No. 34B

Description: This frog is up to 45mm long, and varies from bright leaf-green to pea-green above. There is a paler, thin yellow-green stripe running from the nostril over the eye and tympanum. The entire upper arms, fingers, toes and thighs are bright yellow. The hind edge of the thigh is a purple-brown, often with an iridescent sheen. The belly and throat are cream to yellow. The iris is yellow-orange. The skin is granular above, including the tympanum, and the belly is

Litoria gracilenta   45mm

ia gracilenta

Litoria inermis

also granular. The fingers and the toes nearly fully webbed, and the toe discs are large. The vomerine teeth are between the choanae, and maxillary teeth are also present.

Call: A drawn-out "waaa" or "weee". The males call from reeds and low emergent vegetation. They are usually seen and heard after spring and summer rains.

Habitat: The species is usually found in dense vegetation and reeds associated with water and inundated vegetation. It is also found on wet roads and in flooded ditches, marshes, lagoons and on the wet grassy banks of streams. It can also be attracted to house lights, and often arrives in fruit shops, having travelled in banana bunches.

Similar species: This species is often mistaken for *Litoria chloris*, *Litoria phyllochroa* and *Litoria pearsoniana* from which it can be distinguished by the lack of green colouration on its upper arm, thigh and hand. *Litoria gracilenta* has granular skin on its ear, while *Litoria chloris* does not. *Litoria pearsoniana* and *Litoria phyllochroa* lack obvious webbing between the fingers. *Litoria caerulea* lacks the yellow colouration on hand, thigh and belly.

Floodplain Frog
Litoria inermis
inermis means "unarmed"

Key No. 49C
Description: This is a slender frog, reaching 35mm in length. Its body is grey-brown above, with both lighter and darker indistinct flecks and blotches. The flecks and blotches often mark the positions of warts. There is a stripe through the eye, which is usually faint, and the hind legs are often barred brown. The belly is white, and the backs of the thighs are dark brown with whitish or yellowish blotches. The skin has flat smooth warts above, and it is slightly granular on the belly. The toes are three-quarters webbed, with small toe discs. This frog has both vomerine and maxillary teeth.

Call: A "murk . . . murk . . . murk" sound. The males call from the ground near waterholes.

Habitat: This species is found on flood

Litoria inermis　　　　　　35mm　　　　*Litoria "jervisiensis*

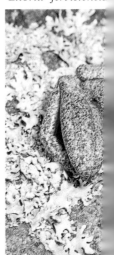

*Litoria
jervisiensis*

with variable darker flecks and spots. There is a dark, broad patch (often indistinct) running down the back. There may be a paler stripe running down this darker patch. A dark stripe runs from the snout, through the eye and tympanum to the base of the arm. The armpit, groin and hind edge of thighs are all red to orange, while the belly is cream or yellow. It has smooth or finely granular skin above. The belly is finely granular. The toes are two-thirds webbed, and the toe discs are large — obviously wider than the toes. There are vomerine teeth between the choanae, and maxillary teeth.

Call: A low-pitched "cree..cree. cree..cree" call. The males call from the water's edge and from the low vegetation nearby. They have been heard calling from August to January.

Habitat: It lives in sclerophyll forest, woodland and shrubland. It is most often seen and heard among reeds and vegetation near streams and inundated areas.

Similar species: This species can be distinguished from *Litoria ewingii, Litoria*

plains, and in woodlands and monsoonal forests. It is usually found after rain.

Similar species: This species is smaller than most others of similar appearance. It can be easily distinguished from *Litoria booroolongensis* and *Litoria lesueuri* by its wartier skin, and the pale bar in front of the eye, where an eye stripe is present. It can be distinguished from *Litoria nasuta* and *Litoria freycineti* by its less distinct body pattern. It can be distinguished from *Litoria latopalmata* by its lack of a white stripe below the eye and tympanum.

Jervis Bay Tree Frog, Heath Frog
Litoria jervisiensis

jervisiensis means "from Jervis Bay", NSW

Key No. 38A

Description: This frog reaches 60mm. It is grey-brown to rich dark brown above,

Litoria "jervisiensis"
Jervis Bay Tree Frog   54mm

Frog   60mm

H. EHMANN

Litoria
latopalmata

Litoria latopalmata
40mm

paraewingi, *Litoria verreauxii, Litoria revelata* and *Litoria dentata* by the lack of a pale stripe below the eye, the extensive red orange colour patches in the groin, armpit and back of thigh, and the large toe pads.

NOTE: There may be two species listed under this name, the Jervis Bay Tree Frog and an undescribed species called the Heath Frog. White *et al.* 1980 (see "References Used") define *L. jervisiensis* as smaller (up to 54mm) with a white glandular stripe below the eye, a yellow (not orange) armpit, and a different call ("two or three high-pitched squeals"), with calling and breeding occurring in winter. This needs to be clarified but for the moment, I have included both species here as *L. jervisiensis.*

Broad-palmed Frog

Litoria latopalmata

latopalmata means "side handed"
Key No. 51B
Description: This frog reaches 40mm. It is a variable species, ranging from pale to dark brown above, with or without darker blotches and variegations. A wide stripe runs from the snout through the eye, over the tympanum and down the flanks, where it breaks up into a series of blotches. The stripe is the same width as the tympanum where it reaches the tympanum. The stripe is broken by a white bar in front of the eye, which curves down below the eye, from where the bar runs to the base of the arm. The backs of the thighs are yellow and dark brown in a reticulated pattern. The belly is white. The skin is smooth, with a few scattered warts, and the belly is granular. The toes have extensive webbing, which is reduced on the fourth and fifth toes. The toe discs are small, not much wider than the toes. The vomerine teeth are

Litoria lesueuri

Litoria lesueuri
70mm

between the choanae, and maxillary teeth are present.

Call: A rapid "yapping" or "quacking" call. The males call from the edges of breeding ponds or from the ground nearby. They have been recorded calling from late spring through summer, after rain.

Habitat: This species can be found well away from water, ranging through all types of forest and open country. During the breeding season, it can be found close to most sources of water— flowing or still, artificial or natural.

Similar species: It can be distinguished from *Litoria lesueuri* by the white bar in front of the eye and the lack of a prominent black and yellow blotched groin (present in *Litoria lesueuri*) and the width of the stripe through the tympanum. Both *Litoria freycineti* and *Litoria nasuta* possess skin folds, lacking in this species. *Litoria inermis* is much wartier, and lacks the white stripe below the eye.

Lesueur's Frog
Litoria lesueuri
lesueuri is after C. Lesueur, the French illustrator

Key No. 51A
Description: This frog reaches 70mm. It ranges from yellow brown to dark brown, and there may be irregular darker blotches present. A dark stripe runs from the snout, through the eye and upper tympanum, to break into a series of blotches on the flanks. The groin is yellow with black blotches. The hind side of the thighs has a reticulated black and yellow pattern. The belly is white. The skin is smooth or finely granular above. The belly is granular, but the throat is smooth. The toes are with well-developed webbing, reduced on fourth and fifth toes, and the discs are small but wider than the toes. There are vomerine teeth between the

choanae, and maxillary teeth.

Call: A soft purring sound. The males call from the ground near the water. They have been heard from August to May.

Habitat: It lives in sclerophyll forest, woodland and associated grassy areas. It is common around rocky flowing creeks, but will breed in still ponds close to these creeks as well.

Similar species: The prominent groin colouration, the toe disc size and the width of the head stripe should distinguish this species from *Litoria booroolongensis*, *Litoria freycineti*, *Litoria latopalmata* and *Litoria brevipalmata*. *Litoria inermis* is easily distinguished from *Litoria lesueuri* because it has a wartier skin.

Rocket Frog
Litoria nasuta
nasuta means "large nosed"

Key No. 50A
Description: This frog reaches 50mm in length, and it is yellow-brown to red-brown on the back, with a series of darker longitudinal warts, ridges and skin folds. The sides are a darker shade of brown. A wide dark stripe runs from the snout, through the eye and tympanum, to break up into a series of blotches along the flanks. This stripe is broken by a white bar in front of the eye and another in front of the arm. The stripe is as broad as the tympanum at this point, and the tympanum is edged with white. A pale stripe runs from below the eye to the base of the arm. The backs of the thighs are yellow with dark brown markings, and the belly is white. The skin is smooth, with irregular skin folds and warts. The belly is granular, but the throat and chest are smooth. The toes are half webbed, except at the fourth and fifth toes, with small toe discs barely

Litoria nasuta

Litoria nasuta
50mm

wider than the toe. The vomerine teeth are between the choanae, and maxillary teeth are present.

Call: A rapid "yapping" or "wik wik wik" sound. The males call from the ground close to water with the onset of spring and summer rains.

Habitat: This species is found in open forests and *Melaleuca* swamps. It is usually associated with streams, ponds, lakes and water-covered grassy areas. It is capable of very long leaps.

Similar species: This species is most similar to *Litoria freycineti*, from which it can be distinguished by the lack of spots on the backs of the thighs. It can be distinguished from *Litoria lesueuri*, *Litoria latopalmata*, *Litoria brevipalmata* and *Litoria booroolongensis* by the rows of longitudinal skin folds and warts on the back. *Litoria inermis* lacks the distinctive body pattern.

Olongburra Tree Frog, Sharp-snouted Reed Frog
Litoria olongburensis
olongburensis means "from Olongburra"

Key No. 46B

Description: This small species reaches 25mm. It is light brown to bright green above, and a dark stripe runs from the nostril, through the eye and tympanum to the flank, and a pale cream stripe runs below this from the eye to the flank. The belly is cream, with brown flecks on the throat. The skin is smooth or finely granular above, and the belly is granular. The toes are moderately webbed, and the fingers are webbed at base. There are vomerine teeth between the choanae, and maxillary teeth.

Call: A "creek...crik" call, similar to *Litoria fallax*. The time of year and the calling position are not known.

Habitat: It lives in emergent vegetation of marshes and swamps. Usually found associated with acidic, tannin-stained water.

Similar species: This species can be distinguished from *Litoria fallax* by its brown throat flecks, and from *Litoria phyllochroa* and *Litoria pearsoniana* by the pale stripe that runs below the dark stripe through the eye, not above it, as in *Litoria phyllochroa*. *Litoria cooloolensis* has a green tympanum, while in *Litoria olongburensis* this is brown.

Litoria paraewingi
40mm

Plain's Tree Frog, Plain's Brown Tree Frog
Litoria paraewingi

paraewingi is similar to the meaning of *ewingii*

Key No. 45B

Description: This up to 40mm long frog is almost identical with Ewing's Tree Frog, *Litoria ewingii*. It is pale fawn or brown on the back, with darker flecks. There is a broad dark band running down the back from between the eyes to the anus. A dark stripe runs from the nostril, through the eye and tympanum, to the flank. The belly is white, and the backs of the thighs are yellow or orange. The skin of the back is smooth with a few tubercles, while the belly is granular. The toes are half webbed, and there are toe discs which are wider than the toes. There are vomerine teeth between the choanae, and maxillary teeth.

Call: The call sounds like "weep . . weep . . weep", but it is slower than in *Litoria ewingii*. The males call from the ground or vegetation by the water's edge. They can also call while floating freely in the water, and have been heard calling throughout the year, but they seem to breed in the spring.

Habitat: They are found in woodland and dry forest, ranging into disturbed land and rural areas. The species can also be found in mountains within its range.

Similar species: It can only be distinguished in the field from *Litoria ewingii* by where it is found (see map), and by its slower calls. It can be distinguished from *Litoria verreauxii* by its lack of dark spots in the groin, and from *Litoria jervisiensis* and *Litoria dentata* by the colour patterns of the thigh and groin, as well as by its smaller toe discs. *Litoria rubella* lacks the broad dorsal stripe and has brown thigh colouration.

Pearson's Tree Frog
Litoria pearsoniana

pearsoniana is after J. Pearson, the Australian zoologist

Key No. 40B

Description: This frog reaches 40mm in length. It is leaf green in colour. A pale white or cream stripe underlined in black runs from the nostril, through the eye and tympanum, down the flanks, where it often breaks up into spots. The under-

Litoria olongburensis

Litoria olongburensis
25mm

lining black stripe widens behind the eye and usually covers the tympanum, taking up most of the width of the side. The groin and hind edge of the thigh are yellowish tan to brick red and the belly is white. The skin is smooth or finely granular above, and granular on the belly. The toes are fully webbed, and the toe discs are well-developed and large. There are vomerine teeth behind the choanae, and maxillary teeth.

Call: A two-part "screech" and "trill" call. The males call from low vegetation or rocks close to the breeding sites. It is apparently a summer breeder.

Habitat: This species has been recorded from rainforest gullies with flowing rocky creeks and streams.

Similar species: This species is very similar to *Litoria phyllochroa* from which it can be distinguished by the broad dark flank stripe (mottled and not taking in most of the side or tympanum in *Litoria phyllochroa*). It can be distinguished from other green tree frogs by its colouration and extent of webbing. *Litoria piperata* is easily distinguished by its mottled or peppered colouration.

Peron's Tree Frog
Litoria peronii
peronii is after F. Peron, the French zoologist

Key No. 35A
Description: This common frog of

Litoria paraewingi

Litoria phyllochroa

the Sydney area reaches 50mm. It varies from grey or dark brown, to cream above, but can change colour rapidly according to temperature, temperament and whether it is night or day. There is usually irregular darker mottling and flecking with a scattering of emerald green flecks. The groin, armpits and hind side of thighs have bold black (or dark brown) and yellow marbling. This colour can sometimes be found on the toes and webbing. The belly is white or cream. The skin is rough with numerous low warts and tubercles, and the belly is granular. The iris is silver with a cross-shaped pupil. The fingers are half-webbed, and the toes are almost fully webbed. Toe discs are large. There are

M MAHONEY

Litoria pearsoniana

Litoria pearsoniana
40mm

vomerine teeth between the choanae, and maxillary teeth are present.

Call: A slow "cackle" descending in inflection and speed as the call progresses. Males call from near the water's edge and from vegetation nearby. It has been heard calling from September to January.

Habitat: This species can be found in most forested habitats but will also forage in grassland and other open areas. It breeds in temporary pools, dams, ditches and inundated areas. It will also attempt to breed in suburban fish ponds.

Similar species: Extremely similar to *Litoria tyleri,* from which it can be distinguished by *Litoria tyleri's* yellow and brown marbling which is only present in the groin and hind side of the thigh, not in the armpits or on the feet. Also, *L. peronii* usually possesses a dark edge to the skin fold over the tympanum. *Litoria rothii* differs in having a red upper half to the eyes, and no green flecking.

Litoria phyllochroa   40mm

Green Leaf Tree Frog
Litoria phyllochroa
phyllochroa means "leaf-coloured"

Key No. 40A, 52C

Description: This frog is up to 40mm long. It is light green to dark olive green above, and it can change colour rapidly. A cream to pale gold stripe runs from the nostril, through the eye and over the tympanum, and down the flanks. This is underlined with a black or dark brown stripe which often breaks up along the flanks to produce a marbled effect with the pale stripe. The armpits, groin, and backs of thighs are all dark red. The belly is white, with occasional darker flecks. The skin is smooth or finely granular above, and granular on the belly. The toes are three-quarters webbed, with large toe discs. There are vomerine teeth behind the choanae, and maxillary teeth. **Note that the tympanum is indistinct in**

Litoria peronii

Litoria peronii

50mm

99

GDEN

some southern specimens.

Call: The call sounds like "erk ... erk ... erk". The males call from the ground near water and from low vegetation, and have been heard calling from October to March, frequently beside running water.

Habitat: It is usually found associated with creek-side vegetation, often sitting on the leaves of *Callicoma* around Sydney during daytime.

Similar species: It is very similar to *Litoria pearsoniana*, from which it can be distinguished by the narrower, broken dark flank band which tends to pass over the top of the tympanum (this stripe is broader and often covers the tympanum in *Litoria pearsoniana*). It can be distinguished from other green frogs by the presence and arrangement of this stripe, as *Litoria pearsoniana* is the only other frog with a broad gold stripe underlined by a narrow dark stripe. *Litoria piperata* lacks this stripe and has mottled or peppered colouring.

Freckled Leaf Tree Frog

Litoria piperata

piperata means "peppered"

Key No. 41B

Description: This frog is 30mm long. It is grey to slate above, with small black dots scattered all over the back and sides, and some green colouration on the head.

There is an obscure dark stripe from the nostril through the eye and tympanum. The belly is pale cream. The skin is smooth, with a few small tubercles above, and the skin is finely granular on the belly. The toes are two-thirds webbed, with large toe discs. There are vomerine teeth behind the choanae, and maxillary teeth.

Call: The call of this species and its breeding habits have not yet been recorded.

Habitat: It is found in vegetation and under rocks, beside streams and ponds in the mountains.

Similar species: The lack of black and yellow marbling in the armpits, groin and backs of thighs distinguish it from *Litoria peronii* and *Litoria tyleri*. It can be distinguished from *Litoria spenceri* by the lack of warty rough skin and its restricted distribution. *Litoria phyllochroa* and *Litoria pearsoniana* lack the "peppered" colouration.

M.PETERSON

H. EHMANN

Litoria piperata

Litoria piperata   30mm

Green or Warty Swamp Frog
Litoria raniformis
raniformis means like *Rana*, a genus of frog more common overseas

Key No. 47B
Description: A large species, up to 85mm. It is olive to bright emerald green above, with irregular gold, brown, black and/or bronze spotting. There is usually a paler green stripe running down the back. A pale stripe runs from the side of the head down the flanks as a fold of skin. The groin and hind edge of the thighs are bright blue; sometimes but rarely with a few small yellow flecks. The belly is white. The skin above is warty and the belly is coarsely granular. The toes are almost fully webbed and the toe

Litoria raniformis   85mm

Litoria raniformis

discs are small — about equal to the toe width. There are vomerine teeth between the choanae, and maxillary teeth.
Call: A growling "waaah..waaah" similar to a motor-boat or-bike. The males call while floating in water among reeds and have been heard from August to April.

Litoria rothii
50mm

Habitat: It is usually found in permanent lagoons, lakes, ponds and dams, especially those with bulrushes and emergent vegetation.

Similar species: The warty back of this species distinguishes it from *Litoria aurea,* and the lack of large yellow spots in the groin and hind edge of the thighs distinguish it from *Litoria castanea.* Its pointed snout, warty skin and blue thigh colour distinguish it from *Litoria burrowsae* (they both occur together in Tasmania).

Whirring Tree Frog
Litoria revelata
revelata means "revealed"

Key No. 38B

Description: This frog reaches 40mm in length. It is cream to red-brown above, with a broad dark brown stripe usually present on the back. A dark stripe runs from the nostril, through the eye and tympanum, to the base of the arm. A pale stripe runs below this. The groin and hind edges of the thighs are orange, with varying amounts of black spotting. The iris is golden. The belly is cream, with brown flecks. The skin is smooth above, with a few raised tubercles, and it is granular on the belly. The toes have medium webbing and large toe discs. There are vomerine teeth behind the choanae, and maxillary teeth.

Call: A whirring call, similar to *Litoria verreauxii.* The males call from the edge of ponds and emergent reeds and grasses nearby. It has been heard calling in late summer and early autumn after rain.

Habitat: It is found in wet and dry sclerophyll forest and coastal swamps. It can be found near dams, ponds and inundated areas.

Similar species: This frog is similar to *Litoria dentata,* but can be distinguished by its lack of red in the upper eye and the presence of black spots in the groin and on the backs of the thighs. This colouration also distinguishes it from *Litoria ewingii, Litoria paraewingi* and *Litoria jervisiensis.* It can be distinguished from *Litoria verreauxii* by the black marbling on the backs of the thighs, which is not present in *Litoria verreauxii. Litoria rubella* lacks a broad stripe down its back.

G.A. & M.M.HOYE

Litoria revelata

Litoria revelata
40mm

Litoria rothii

Roth's Tree Frog
Litoria rothii
rothii is after L. Roth, the collector of these frogs

Key No. 36B
Description: This up to 50mm long frog varies from grey to pale brown on its back, with some darker mottling. There are no green speckles or flecks on the back, and the belly is white. The armpits and the base of the arms are usually black. The backs of the thighs are a black blotch or stripe over a bright yellow patch. The upper half of the eye is either bright red or rust red. The skin of the back is rough, with low warts and tubercles, and the belly is granular. The fingers are about half webbed, and the toes are almost fully webbed, with large toe discs. There are vomerine teeth between the choanae, and maxillary teeth.

Call: A chuckling or cackling call, similar to *Litoria peronii*, only slower. The males call through the summer months, particularly after summer rain. They are usually found calling from the ground or from low vegetation near water.

Habitat: The frog is common around trees and other cover close to water. In the daytime, it is often found sheltering under bark or in hollow branches. It can also be found some distance from permanent water, relying on temporary puddles, ditches and claypans for breeding.

Similar species: This species can be eas-

Litoria rubella 35mm

ily distinguished from *Litoria peronii* and *Litoria tyleri* by its lack of green flecking on the back, and by its "two-toned" eyes. *Litoria dentata* also has red in the upper half of its eyes, but it has a broad stripe down the back, which is absent in *Litoria rothii*.

Desert Tree Frog
Litoria rubella
rubella means "reddish"

Key No. 39C

Litoria rubella

Description: This small, short-legged frog of up to 35mm is the most widely distributed in Australia, being found in all states except Victoria and Tasmania. It varies from grey, red-brown, to fawn above, with occasional darker flecking. A dark band runs from the nostril through the eye and tympanum and down the flanks. The groin is usually yellow, while the hind edges of the thighs are brown with white flecking. The belly is white to yellowish. The skin is smooth to finely granular above, and granular on the belly. The toes are about two-thirds webbed, with large toe discs. There are vomerine teeth behind the choanae, and maxillary teeth.

Call: A harsh buzz. The males call from the ground and low vegetation close to water. Calling usually coincides with summer rains.

Habitat: This species can be found in most available habitats, but is mainly restricted to the west of the Great

Dividing Range in southern Australia. They frequently take up residence around human habitation, and many may be found sheltering together by day. **Similar species**: Its appearance and distribution are likely to distinguish it from most other south-east Australian frogs. It lacks the red in the upper half of the eyes which is found in *Litoria dentata* and *Litoria rothii*. It lacks any dark band on the back, such as is found in *Litoria ewingii*, *Litoria paraewingi*, *Litoria verreauxii*, *Litoria jervisiensis* and *Litoria revelata*.

Litoria spenceri   45mm

Litoria spenceri

Spotted Tree Frog
Litoria spenceri
spenceri is after W. Spencer, the Australian zoologist

Key No. 52A
Description: This frog is up to 45mm long. It is grey to olive green above, with irregular spotting and marbling. The belly is white to yellowish. The skin is granular to rough above, and granular on the belly. The toes are fully webbed, with toe discs only a little wider than the toes. There are vomerine teeth behind the choanae, and maxillary teeth.
Call: A harsh "warrrk....cruk..cruk.. cruk..cruk..cruk" call. The males call from the ground or low vegetation near mountain streams. Calls have been heard in November and December.
Habitat: It lives in wet and dry sclerophyll, usually associated with flowing rocky streams and creeks. This species has declined in recent years.
Similar species: Due to the distinctive colouration and distribution, it is unlikely to be confused with any other frog. The southern forms of *Litoria phyllochroa* (which may lack a distinct tympanum) do not have a warty granular skin.

New England Tree Frog
Litoria subglandulosa
subglandulosa means "below glandular"

Key No. 52B
Description: This frog reaches 50mm long. The colour ranges from green to olive brown above, and the sides usually remain green. A golden stripe runs from the nostrils, through the eyes and above the indistinct tympanum, and down the flanks. A broad, dark stripe runs beneath it. The backs of the thighs are red-brown. The skin is smooth or finely granular above, and granular on the belly. The toes are almost fully webbed, with large toe discs. There are vomerine teeth behind the choanae, and maxillary teeth.
Call: An "orak..orak..orak" call with varying speed and volume. It has been heard calling from October to November.
Habitat: It is found in small streams on the New England Ranges.

G.A.HOYE

Litoria subglandulosa
50mm

Similar species: This species is most similar to *Litoria citropa*, from which it can be easily picked out by the hidden or indistinct tympanum, which is distinct in *Litoria citropa*. Apart from its distribution, it can be told apart from specimens of *Litoria phyllochroa*, which may have an indistinct tympanum (more common

Litoria subglandulosa

Litoria tyleri
50mm

Litoria tyleri

Litoria verreauxii

Litoria verreauxii
30mm

in Victoria), by the broad dark stripe which underlies the gold stripe. In *Litoria phyllochroa*, the dark stripe is narrow.

Tyler's Tree Frog
Litoria tyleri
tyleri is after M. Tyler, the Australian zoologist

Key No. 35B
Description: This frog is up to 50mm long and grey-brown to fawn above, with scattered darker flecks. It also has a scattering of emerald green flecks. The armpits, groin and hind edge of the thighs are bright yellow. The groin and backs of thighs have brown marbling, frequently running parallel to the edge of the thighs. The belly is white, or yellowish-white. The skin is finely granular above, and granular on the belly. The iris is golden, with a cross-shaped pupil. The fingers are half webbed and the toes are almost fully webbed, with large toe discs. There are vomerine teeth between the choanae, and maxillary teeth.
Call: A short, rattling sound, similar to that of *Litoria peronii* but without the descending inflection. The males call from rocks and logs on the ground close to water and from associated vegetation above it, from spring to late summer, usually after rain.
Habitat: It lives in wet and dry sclerophyll forest and adjacent areas, especially near swamps, dams, creekside ponds and reedy lagoons.
Similar species: This species is very similar to *Litoria peronii*, from which it can be distinguished by the lack of dark spots in the armpits and the brown and yellow marbling. As well, *Litoria peronii* has a silver iris and usually a dark edging over the tympanum. The groin and thigh colouration should distinguish it from any other similar species. *Litoria rothii*

differs in having a red upper half to the eyes, and no green flecking.

Verreaux's Tree Frog
Litoria verreauxii
verreauxii is after J. Verreaux, the French zoologist

Key No. 45A
Description: This is a variable frog, reaching 30mm in length. Lowland forms range from light brown to red-brown above, with scattered darker flecks. A darker broad patch or band usually runs from between the eyes, all the way down the back. This patch is divided by a paler stripe running down the backbone. A dark stripe runs from the nostrils, through the eyes, and covering the tympanum, to the base of the arms and down the flanks, where it becomes a series of blotches. A whitish stripe runs below this, from the eyes to the base of the arms. The groin, front and hind sides of the thighs are yellow through to red-orange. The groin has black spots or blotches. Mountain forms are similar to the lowland forms, but the patches on the back and some of the side patches are green with brown edges. The bellies of both forms are whitish. The skin is smooth or finely granular above, with a few low warts. The belly is granular. The toes are half webbed, and the toe discs are small, only a little wider than the toes. There are vomerine teeth between the choanae, and maxillary teeth.
Call: A "weep..weep..weep..weep" sound. The males call from the ground, or low vegetation bordering the breeding pond, all year round.
Habitat: It can be found in most habitats, from sea level to the mountains and forests to grasslands.
Similar species: This species can be distinguished from *Litoria ewingii* and

Litoria paraewingi by its divided patch on the back and the spotted groin. It can be distinguished from other similar species *Litoria jervisiensis, Litoria dentata* and *Litoria revelata* by a combination of its reduced toe disc size, reduced webbing, spots in the groin, lack of colour in the armpits and divided back marking. *Litoria rubella* has no broad stripe down the back.

FAMILY: BUFONIDAE

This family has genera and species found throughout most of the world. The family does not occur naturally in Australia — our one species was introduced in 1935 to control cane beetles, but this was not successful. Instead, the toads established themselves and became a pest in their own right.

GENUS

Bufo

This genus has a single representative species in Australia. It is found in Queensland, the Northern Territory and New South Wales. It has fully webbed toes, enlarged parotoid glands and horizontal pupils. The eggs are laid in strings.
Bufo *means "toad".*

Bufo marinus   150mm

Cane Toad
Bufo marinus
marinus means "marine"

Key No. 2C
Description: A very large amphibian, up to 150mm long, it is grey, olive, brown or red-brown above. Small specimens have

Bufo marinus

darker patches and markings. It is whitish or with a yellow tinge on the belly, with brown speckles and flecks. The skin is rough and warty above, with prominent parotoid glands behind the eyes, and the belly is granular. The toes are fully webbed. There are no vomerine or maxillary teeth.

Call: A "purring" sound, similar to a higher pitched telephone dial tone. The males call at the water's edge during summer rains.

Habitat: It is found in most habitats within its range. It can breed in brackish water and has also been found in mangroves. It readily makes its home around human habitation, feeding on insects attracted to outside lights and breeding in urban fish ponds.

Many are seen on the roads at night. It is an extremely tough and adaptable species, and appears quickly to outnumber native frogs when it colonises a new area. It is probable that it competes with some native species for resources. It is also poisonous at all stages of its development — egg, tadpole and adult.

Similar species: Due to its size and enlarged parotoid glands, this species is difficult to confuse with any other. Its eggs are strung together like a necklace.

Special Note: Handle the Cane Toad with care, as the milky fluid from its neck glands is quite dangerous. You should wash your hands after handling any frog or toad, but this is especially important after handling Cane Toads.

REFERENCES USED

Barker, J. and Grigg, G., 1977, *A Field Guide to Australian Frogs*, Rigby, Adelaide.

British Museum (Natural History), 1980, *Man's Place in Evolution*, Cambridge University Press.

Brook, A.J., 1980, "The breeding seasons of frogs in Victoria and Tasmania", *Victorian Naturalist*, **97**, 6-11.

Brook, A.J., 1989, *The Zoogeography of Victorian Anura*, M.Sc. Thesis, University of Melbourne.

Clyne, D., 1969, *Australian Frogs*, Lansdowne Press, Melbourne.

Cogger, H.G., 1960, *The Frogs of New South Wales*, Government Printer, Sydney.

Cogger, H.G., 1992, *Reptiles and Amphibians of Australia*, 5th edition, Reed Books, Sydney.

Davies, M. & Littlejohn, M.J., 1986, "Frogs of the genus *Uperoleia* Gray (Anura: Leptodactylidae) in South-eastern Australia". *Transactions of the Royal Society of South Australia*, 110, 111-143.

Davies, M., McDonald, K. R., and Corben, C. J., 1986, "The genus *Uperoleia* Gray (Anura: Leptodactylidae) in Queensland, Australia". *Proceedings of the Royal Society of Victoria*, **98**(4), 147-188.

\Grigg, G. & Barker, J., 1973, *Frog Calls of South East Australia*, (cassette tape & information booklet), Zoology Dept, University of Sydney.

Hero, J.M. & Littlejohn, M.J., 1991, *A field guide to the frogs of Victoria*, draft, Dept of Zoology, University of Melbourne.

Hero, J. M., Littlejohn M., and Marantelli G., 1991, *Frogwatch Field Guide to Victorian Frogs*, Dept of Conservation and Environment, Victoria.

Ingram, G., Corben, C.J. & Hosmer, 1982, "*Litoria revelata*, a new `species of tree frog from eastern Australia", *Memoirs of the Queensland Museum*, **20**(3), 635-637.

Littlejohn, M.J., 1963, "Frogs of the Melbourne Area", *Victorian Naturalist*, **79**: 296-304.

Littlejohn, M.J., 1981, "The Amphibia of mesic southern Australia: a zoogeographic perspective". In: A.Keast (ed), *Ecological Biogeography of Australia*. Junk, The Hague.

Martin, A.A. & Littlejohn, M.J., 1982, "Tasmanian Amphibians", *Fauna of Tasmania Handbook No. 6*., University of Tasmania.

Moore, J.A., 1961, "The Frogs of Eastern New South Wales", *Bulletin of the American Museum of Natural History*, **121**(3), 151-385.

Tyler, M.J., 1978, *Amphibians of South Australia*, Government Printer, Adelaide.

Tyler, M.J., 1989, *Australian Frogs*, Viking O'Neil, Penguin Books, Ringwood.

Tyler, M.J. & Davies, M., 1985, "A new species of *Litoria* (Anura: Hylidae) from New South Wales, Australia.", *Copeia*, 1985 (1), 145-149.

White, A. W., Whitford D. and Watson, G. F., 1980, "Redescription of the Jervis Bay Tree Frog *Litoria jervisiensis* (Anura: Hylidae) with notes on the identity of Krefft's Frog (*Litoria kreffti*)", *The Australian Zoologist*, **20**, part III.

GLOSSARY

Arboreal: Tree-dwelling or tree-climbing.

Choanae: The internal openings of the nostrils and nasal passages in the roof of the mouth.

Cloaca: The common chamber into which both reproductive and waste products enter before passing out of the body via the cloacal opening or vent.

Coccygeal gland: An enlarged lump producing skin secretions, located over the side of the rump.

Discs: Expanded ends on the fingers and toes, usually with structures and secretions that allow the animal to "stick" by surface tension to smooth surfaces.

Dorso-lateral: Situated between the back and side of the animal, as in a dorso-lateral fold.

Dorsal: Refers to the back or upper surface of the body.

Ectothermic: Animals in which body heat is dependent on external heat. Ectothermic animals are often incorrectly called "cold-blooded".

Flank: The sides, between the front and back limbs.

Fringed toes: Toes with a thin, transparent "skirt" of skin around the side. This fringe may be folded around the toe, so look carefully.

Granular: Presenting a "cobbled" or "smooth gravelled" appearance to the surface. The size of the individual "cobbles" determines whether the surface is finely granular or coarsely granular.

Groin: The angle formed where the hind limb joins the body.

Inguinal gland: An enlarged lump producing skin secretions, situated in or near the groin.

Lateral: Refers to the side of the body or limb.

Maxillary teeth: The row of teeth around the upper jaw.

Metamorph: A frog immediately after the tadpole stage.

Metamorphosis: The change in body form and function from larva to adult (e.g. tadpole to frog).

Metacarpal tubercle: A lump or projection extending from the underside of the hand, often also called a palmar tubercle.

Metatarsal tubercle: A lump or projection extending from the underside of the foot.

Palmar tubercle: See "Metacarpal tubercle" above.

Parotoid gland: An enlarged lump producing skin secretions, situated behind the eyes in the shoulder region.

Posterior: The hind end of the body or structure.

Reticulated: Forming a network pattern.

Sexual dimorphism: Differences in shape, colour or size between males and females of the same species.

Tibial gland: An enlarged lump producing skin secretions, located on the upper surface of the lower legs (the tibial section of the legs).

Tympanum: The eardrum — lying flush with the skin in frogs.

Vent: The hole through which waste products and reproductive cells pass out of the body; the external opening of the cloaca.

Ventral: Refers to the belly and/or undersurface of the body.

Vertebral: Relating to the spine or "backbone" of the body.

Vomerine teeth: Teeth attached to a bone (the vomer) in the roof of the mouth near the choanae.

Wallum: Heath, scrub and woodland on sand dunes and sand plains between Fraser Island and Sydney.

Webbing: A flat skin membrane stretching between the toes (or fingers). The extent of webbing is described by terms such as "half webbed", meaning the webbing extends halfway up the length of the toe.

INDEX